Automated Vehicles: Multi-Agent Safety

Automated Vehicles: Multi-Agent Safety

JUAN R. PIMENTEL

Professor of Computer Engineering
Kettering University

SAE INTERNATIONAL®

Warrendale, Pennsylvania, USA

INTERNATIONAL®

400 Commonwealth Drive
Warrendale, PA 15096-0001 USA
E-mail: CustomerService@sae.org
Phone: 877-606-7323 (inside USA and Canada)
Fax: 776-4970 (outside USA)

Library of Congress Catalog Number 2019931390
SAE Order Number PT-204
http://dx.doi.org/10.4271/pt-204

ISBN-Print 978-0-7680-0219-5
ISBN-MediaTech 978-0-7680-0220-1
ISBN-prc 978-0-7680-0223-2
ISBN-epub 978-0-7680-0224-9
ISBN-HTML 978-0-7680-0225-6

To purchase bulk quantities, please contact: SAE Customer Service

E-mail: CustomerService@sae.org
Phone: 877-606-7323 (inside USA and Canada)
Fax: 776-4970 (outside USA)

Visit the SAE International Bookstore at books.sae.org

contents

CHAPTER 8

Hardware-in-the-Loop (HIL) Implementation and Validation of SAE Level 2 Autonomous Vehicle with Subsystem Fault Tolerant Fallback Performance for Takeover Scenarios 93

CHAPTER 9

Vehicle-GIS Assistant Driving System for Real-time Safety Speed Warning on Mountain Roads 127

Introduction

Safety has been ranked as the number one concern for the acceptance and adoption of automated vehicles (AVs) and understandably so since safety has some of the most complex requirements in the development of self-driving vehicles. The recent fatal accident involving self-driving vehicles has made it clear that safety is paramount to the acceptance, testing, verification, and deployment of AVs. As one of the main categories of AV safety, multi-agent safety is the one that most closely resembles the safety of vehicles when driven by a human; this means dealing with other vehicles, pedestrians, bicyclists, and any other object encountered while driving. There are a number of outstanding safety issues that this book will help clarify such as what exactly is multi-agent safety and how to measure the level of safety of AVs, for example, is the number (in millions) of miles driven a good indicator? And if so, under what circumstances or conditions?

Index Terms

Automated vehicles

Self-driving vehicles

Autonomous vehicles

Safety

SOTIF

Multi-agent safety

Functional safety

Safety measurement

Safety evaluation

I.1 Introduction

Automated vehicles (AVs), also called autonomous or self-driving vehicles, have the potential to reduce accidents, help with the environment, reduce congestion, help the elderly and other disadvantaged populations, and produce other societal benefits [1, 2, 3, 4, 5]. However, the touted advantages of autonomous vehicles and those including latest advanced driver assistance system (ADAS) features are turning out difficult to sell to the public than many manufacturers and tier 1s have anticipated. Much of the early euphoria of self-driving vehicles is diminishing in the wake of some recent accidents involving AVs with varying degrees of automation. A recent online marketplace for buying and selling cars found 69% of respondents are scared of autonomous automobiles. It is also found that these people found technology in cars helpful (58%), but only 12% said ADAS and infotainment features were a "must-have." The survey asked more than 1,000 respondents from across the United States geographically and across age groups, although it should be noted the biggest group was 60+ years old.* Accidents involving self-driving vehicles are inevitable; as it is the case with other industries, accidents have happened and will happen no matter the efforts made to avoid them. The National Transportation Safety Board (NTSB) has issued a report† on a Tesla accident on May 7, 2016, and two preliminary reports on an Uber accident‡ on March 18, 2018, and a Tesla accident§ on March 23, 2018. After analyzing these reports, what is disturbing are the details associated with these accidents which indicate that as an industry, we may need to go back to safety 101.¶

Regarding the aforementioned accidents, it would not be so bad if the safety systems of the vehicles in question were designed and functioning properly according to their stated automation level. In the case of the Tesla accident in Florida, the vehicle failed to activate the forward collision warning (FCW) system and automatic emergency braking (AEB). In the case of the Uber accident, emergency braking maneuvers were not enabled while the vehicle was under computer control, to reduce the potential for erratic vehicle behavior, and the system relied on the vehicle operator for safety. In the Tesla accident in California, the vehicle failed to detect a damaged crash attenuator and hit it at a speed of about 71 mph.

After recent incidents and mishaps involving AVs such as those described above, it is clear that there is much room for improvement

* http://analysis.tu-auto.com/autonomous-car/shifting-public-acceptance-autonomous-tech?NL=TU-001&Issue=TU-001_20180723_TU-001_235&sfvc4enews=42&cl=article_2_2.

† https://www.ntsb.gov/investigations/AccidentReports/Reports/HAR1702.pdf.

‡ https://www.ntsb.gov/investigations/AccidentReports/Reports/HWY18MH010-prelim.pdf.

§ https://ntsb.gov/investigations/AccidentReports/Reports/HWY18FH011-preliminary.pdf.

¶ https://www.eetimes.com/document.asp?doc_id=1333446&_mc=RSS_EET_EDT&utm_source=newsletter&utm_campaign=link&utm_medium=EETimesWeekInReview-20180721.

not only by manufacturers but also by government regulations, researchers, the general public, and other stakeholders. Over the past few months, the media has been full of headlines such as "How Safe Is Driverless Car Technology, Really?," "Autonomous Cars: How Safe Is Safe Enough?," and "How safe should we expect self-driving cars to be?." In addition, some industry analysts and safety experts are offering advice to tech and automotive companies to reconsider their safety programs. There is also some agreement that "the self-driving car industry's reputation has suffered a setback," and the question is how to fix it*. It appears that AV companies are much more stringent when using semiconductor devices and EDA tools demanding that they conform to ISO 26262 than using the same yardstick for their own safety-critical designs.

So what is there to do? Safety is not new; at least for the last 60 years, it has been successfully applied in several industries such as nuclear, avionics, process control, automotive, and others. What is unique and special about the safety of self-driving vehicles? What should be the emphasis for a more effective AV safety program? What are the roles of governments, standards, testing, verification, validation, and sound safety engineering efforts? Addressing issues regarding autonomous vehicle safety is challenging [7]. Currently, as an industry, we just do not fully understand the nature of self-driving vehicle safety and how to design safe AVs. For example, there is little discussion on ways to estimate, analyze, compute, or measure the level of safety of an AV design or AVs. We need to begin by fully characterizing it, and this book series is an effort in this direction.

Some manufacturers such as Waymo cite their recent milestone of 8 million miles driven on public roads as a measure of the safety achieved by their self-driving vehicles.† However, it is not clear how a certain number of millions of miles driven contribute to the safety level of self-driving vehicles. Some industry analysts believe that policymakers and city officials overseeing infrastructure will be the most important players in reshaping the self-driving vehicle safety landscape. For example, the National Highway Traffic Safety Administration (NHTSA) has issued a voluntary guidance whose purpose is to help designers of automated driving systems (ADSs) analyze, identify, and resolve safety considerations prior to deployment using their own, industry, and other best practices.‡ It outlines 12 safety elements, which the agency believes represent the consensus across the industry, that are generally considered to be the most salient design aspects to consider and address when developing, testing, and deploying ADSs on public roadways. Within each safety design element, entities are encouraged to consider and document their use of industry standards, best practices, company

---

* https://www.eetimes.com/document.asp?doc_id=1333446&_mc=RSS_EET_EDT&utm_source=newsletter&utm_campaign=link&utm_medium=EETimesWeekInReview-20180721.

† https://www.theverge.com/2018/7/20/17595968/waymo-self-driving-cars-8-million-miles-testing.

‡ https://www.nhtsa.gov/sites/nhtsa.dot.gov/files/documents/13069a-ads2.0_090617_v9a_tag.pdf.

policies, or other methods they have employed to provide for increased system safety in real-world conditions. The 12 safety design elements apply to both ADS original equipment and replacement equipment or updates (including software updates/upgrades) to ADSs. However the NHTSA guidance is not specific enough to help manufacturers design effective safety mechanisms to reduce risk.

I.2 Characterizing the Safety of Automated Vehicles

How different is the concept or notion of safety in self-driving vehicles when compared to that used in other industries such as aviation, process control, and automotive? While the fundamental concepts are the same, the safety of self-driving vehicles has specific attributes that are different or not present in the safety of other industries. In this section we briefly discuss these attributes. When compared to the safety of traditional industries such as avionics, process control, and automotive, there are specific attributes pertaining to the safety of self-driving vehicles that we discuss next [23].

I.2.1 Performance Degradation

Traditional safety is based on faults and failures of mostly hardware components, and this is referred to as the reliability approach to safety [9]. In contrast, accidents involving self-driving vehicles might happen even if no hardware device fails, but rather a performance degradation of some of its functions or intended functionality occurs. Addressing safety issues for these situations is referred to as Safety of The Intended Functionality (SOTIF), and it is a fairly new concept as applied to the safety of self-driving vehicles [10]. Thus, some failures are due to performance degradation of self-driving vehicle components, typically involving higher levels of processing or higher levels of automation, for example, service failures. This definition of failure goes beyond that which is defined in the standard ISO 26262; however, it is compatible with other safety frameworks such as [8], STPA [11, 12, 13, 14, 15], or other real-time distributed systems [16]. One example of this understanding of the concept of safety is the failure of a vehicle detection system where the perception system provides missed detections (i.e., false negatives) or spurious detections (i.e., false positives). This could happen because the processing of environmental data is highly complex and the object detection function is subject to errors and impairments, particularly in bad weather or in night conditions when the visibility is poor. Either one of these failures could be catastrophic and could result in an accident or harm. Another example is a radio detection and ranging (RADAR) system correctly detecting objects only when the objects are moving, thus missing static objects because of limits on its performance. Thus, failure occurs in a degraded performance scenario.

I.2.2 Focus on Software

It is well known that the amount of software in a vehicle continues a rapidly increasing trend that started with the development of by-wire systems. Much of the functionality of a self-driving vehicle is implemented in software, and thus it is important to view the perception system as a set of software servers, each providing services to the rest of the system. Therefore, one can refer to these various functionalities as a vehicle detection server, a pedestrian detection server, a road detection server, etc. The software in self-driving vehicles is much larger in size and scope compared to traditional industries; thus, there should be a focus on the safety of the software. As noted, a failure can occur if the software services deviate from the correct services, and this could lead to safety hazards and safety risks. Ultimately, the overall safety of a self-driving vehicle will be dictated by the safety of its software [9, 12].

I.2.3 Non-Deterministic Perception System

In the absence of hardware faults, the perception systems of traditional industries are mostly deterministic in nature. For example, sensing the intake manifold pressure or engine speed in automotive systems is deterministic[4]. In contrast, the perception systems of self-driving vehicles are non-deterministic, leading to a high level of false positives and false negatives when their performance deteriorates to the point that service failures cannot be avoided. The non-deterministic aspect of the perception system stems from the fact that one never knows when its performance will deteriorate to the point where failures begin to appear in the services delivered by the system. Thus, the services provided by the perception system are subject to random failures, for example, when the weather deteriorates or when the system makes detection errors.

I.2.4 Perception System Complexity

Sensing elementary physical phenomena such as temperature or pressure is relatively simple, involving just some deterministic sensors, some electronics, and communications. In contrast, sensing or detecting man-made entities or constructs such as another vehicle, a road boundary, a city street, or a street intersection is complex because of the lack of structure of what is being sensed or perceived. The implication of the complexity of the perception system is that it is prone to errors, which degrades the performance or safety of the overall vehicle.

I.2.5 Overall System Complexity

In addition to its perception system, a self-driving vehicle also includes localization and mapping, planning and control, and actuation resulting in a highly complex system. One of the main issues with system complexity is that it makes testing for safety challenging, particularly if machine learning techniques are used, as it makes the design opaque to humans. This makes tracing the design and the test plans to the requirements problematic, since there is no human-understandable design that can be used for verification and testing [17]. In addition, it is known that when the system is complex, the system safety is affected by interacting complexity and tight coupling [9]. Another aspect of system complexity is that the autonomous vehicle operates in a complex external environment, and there are safety hazards due to events outside the domain of the autonomous vehicle, for example, from other vehicles (whether self-driving or not). Thus, the safety attributes of a self-driving vehicle are significantly different from those in other industries such as avionics, process control, and automotive.

In addition to its attributes, what are the various types of safety that encompass the overall safety of self-driving vehicles? As noted, the safety of self-driving vehicles is complex and differs from that of other industries such as avionics, process control, and automotive. On the one hand, there are safety commonalities such as the safety that involves component failures, which is the subject of so-called functional safety, and the safety involving components whose failure rates are well understood because they are proven in use, that is, in actual operation. On the other hand, there are two types of safety that are not prevalent in the avionics, process control, and automotive industries, and these include *SOTIF* and *multi-agent safety*. Thus, the types of safety that characterize the safety of self-driving vehicles include (1) traditional functional safety as defined by ISO 26262, (2) SOTIF, and (3) multi-agent safety [23]. Feth et al. also emphasize that safety assurance is a concern because established safety engineering standards and methodologies are currently not sufficient [22]. They also conclude that there are three types of safety that characterize the safety of self-driving vehicles: (1) traditional functional safety, (2) SOTIF which they assume are due to *functional insufficiencies*, and (3) multi-agent safety related to safe driving behaviors which are abstracted from technological challenges of situation awareness. Furthermore, they elaborate the fundamental safety engineering steps that are necessary to create safe vehicle of higher automation levels while mapping these steps to the guidance presently available in existing (e.g., ISO 26262) and upcoming (e.g., ISO PAS 21448) standards. Functional safety is a well-understood area which is guided by a number of international standards such as IEC 61508 [18], IEC 61511 [19], and ISO 26262, and there are a large number of papers and publications on this topic. However, it is noted that ISO 26262 does not cover AVs; thus, its application should be done with great care. In the following, we characterize the safety category of multi-agent safety.

I.3 Multi-agent Safety

In terms of safety, designing an autonomous flying machine is much easier than designing a self-driving vehicle, because currently there is simply much less in the sky to deal with in regard to other flying machines or agents, in general. Unlike avionics, process control, and automotive, accidents in autonomous vehicles can happen due to actions of agents such as other vehicles, pedestrians, bicyclists, and in general any object that is external to the vehicle. Addressing safety issues in this type of environment is known as multi-agent safety. An example of a specific multi-agent situation that might involve a safety hazard is that of a cut-in that happens when a vehicle changes lanes and positions itself ahead of the self-driving vehicle on a highway. For this case, the set of control actions (CA) are brake, turn left, and turn right, which can lead to accidents if not performed safely. As is the case for SOTIF, much more research is needed to fully characterize multi-agent safety [20, 21].

Currently we lack a sufficient number of formal models and theories to address safety of AVs from the viewpoint of multi-agent safety. One such formal model is detailed in [20] and in this section we summarize it and extract some of its most important elements. A model-based approach for multi-agent safety of AVs is important in that it complements the *functional safety* approach which is probabilistic and risk based. In the context of multi-agent safety, an important question is whether we can *guarantee safety*? If so, under what conditions? Is it possible to guarantee zero accidents and achieve *absolute safety*? This is impossible as shown in the situation depicted in Figure 1(a) since the vehicle at the center is at the mercy of the surrounding four vehicles. Likewise, one can envision situations where an accident occurs solely due to the blame of other agents. This is illustrated in Figure 1(b) and (c) if vehicle c_1 performs a cut-in that is not safe. The model in [20] introduces the concept of *responsibility-sensitive safety* (*RSS*) which formalizes the notion of *accident blame*, it is interpretable and explainable, and it incorporates a sense of *responsibility* into the actions of a *robotic agent*.

Responsibility-Sensitive Safety. The paper [20] argues that a crucial aspect that is missing from the absolute safety concept is the non-symmetry of most accidents—it is usually one of the drivers who is responsible for a crash and is to be *blamed*. For the example in Figure 1 (b) and (c), vehicle c_0 is not to be blamed if c_1, for example, suddenly drives into it. This paper formalizes the fact that considering its lack of

FIGURE 1 Multi-agent driving situations, (a) central vehicle cannot ensure safety, (b) front and back corridors, (c) cut-in situation.

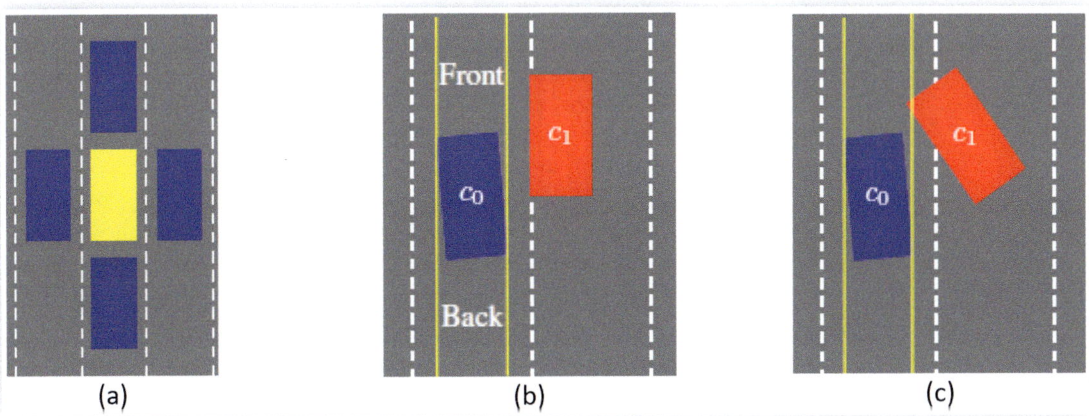

responsibility, a behavior of staying in its own lane can be considered safe. In order to do that, the authors develop a formal concept of *accident blame*, which captures the common sense behind safe driving. Intuition is gained with an informal examination of one of the most basic responsibility concepts that is later formally developed. The authors consider the simple case of two cars c_f and c_r, driving at the same speed, one behind the other, along a straight road. The following is a typical assertion made in the paper: "Assume c_f, the car at the front, suddenly brakes because of an obstacle appearing on the road, and manages to avoid it. Unfortunately, c_r did not keep enough of a distance from c_f, is not able to respond in time, and crashes into c_f's rear side. It is clear that the blame is on c_r; it is the responsibility of the rear car to keep safe distance from the front car, and to be ready for unexpected, yet reasonable, braking." The authors then formally consider a much broader family of scenarios: driving in a multilane road, where cars can freely change lanes, cut into other cars' paths, drive at different speeds, and so on. To simplify the discussion, a straight road on a planar surface is assumed, where the lateral and longitudinal axes are the x and y axes, respectively.

In the following we provide some modeling elements and definitions extracted from [20]. Agents play a nonsymmetrical role in an accident where typically only one of the agents is responsible for the accident and therefore is to be *blamed* for it. The model introduces the concept of *cautious driving*, useful for situations under limited sensing conditions, when not all agents are always visible (e.g., due to occlusions of kids behind a parking vehicle). An action taken by a car *c* is *absolutely safe* if no accident can follow it at some future time. What kind of actions or behaviors (e.g., staying in its own lane) can be considered safe? It is clear that vehicle safety in a multi-agent environment has to take into account the dynamics of the vehicles (ego and other), pedestrians, bicyclists, and other objects. For vehicle dynamic considerations, it is important to adopt an appropriate coordinate system, such as that defined by ISO 8855:2011.

The *blame time* of an accident is the earliest time preceding the accident in which there was an intersection between one of the cars and the other's corridor or lane, and the longitudinal distance was not safe. The *exposure time* of an object is the first time in which we see it. Assume that at the exposure time or after it, car c_1 was driving at speed $v > v_{limit}$, and c_0 wasn't doing so. Then, the blame is only on c_1. We say that c_1 is *blamed due to unreasonable speed*. The *accident-with-pedestrian* blame is always on the car, unless one of the following three situations holds: (a) the car hits the pedestrian with the car's side, and the lateral velocity of the car is smaller than w.r.t. the direction of the hit, (b) the pedestrian's velocity at the exposure time or later was larger than v_{limit}, and (c) the car is at complete stop. The *default emergency policy* (DEP) is to apply maximum braking power and maximum heading change toward 0 heading with respect to the lane. A *state s* is *safe* if performing DEP starting from it will not lead to an accident of our blame. Suppose we are currently at state s_0. A *command a is cautious* if the next state, s_1, will be safe with respect to a set *A* of possible commands that other vehicles might perform now. Let *S* denote the domain of sensing state and let *X* be the domain of raw sensor and mapping data. A *sensing system* is a function $X \rightarrow S$.

I.3.1 Efficiently Validated Conditions for Responsibility-Sensitive Safety

To be useful, definitions must be accompanied by efficient ways of successfully verifying that can never be blamed for an accident. A *brute-force* approach that will check all possible future outcomes would be computationally impossible. The authors then accompany the RSS definitions with computationally efficient methods to validate them. With additional definitions, assumptions, and mathematical methods such as *induction*, the authors develop checks and algorithms to support some *RSS* situations such as calculating the minimal safe longitudinal distance to a front vehicle and an algorithm for checking the possibility of a non-safe cut-in [20]. These are summarized below.

I.3.2 Calculation of the Minimal Safe Longitudinal Distance

A longitudinal distance between a car c_r and another car c_f that is in c_r's frontal corridor is *safe* with respect to a response time ρ if for any braking command a, $|a| < a_{max,brake}$, performed by cf, if cr will apply its maximal brake from time ρ until a full stop then it won't collide with cf.

This *safe longitudinal distance* can be calculated in terms of the following parameters:

1. $a_{max,brake}$, $a_{max,accel}$: Maximum braking and acceleration commands
2. vr, vf: Longitudinal velocities of the rear and front cars
3. lf, lr: Lengths of the cars
4. $v_{\rho,max} = vr + \rho \times a_{max,accel}$
5. ρ: Response time

I.3.3 Safe Cut-In of the Ego Vehicle

What are the checks (conditions, requirements) so that there can be a safe cut-in of the ego vehicle? It is possible to develop algorithms for checking the possibility of a non-safe cut-in in terms of the following parameters:

1. $\hat{c}$, c: Ego and another vehicle
2. $[0, t_{brake}]$: Time interval in consideration
3. $\hat{c}_{diag}$: Length of a diagonal of a minimal rectangle bounding $\hat{c}$
4. $c_{length(t)}$: Longitudinal "span" of c in time t
5. $L(t) = (\hat{c}_{diag} + c_{length(t)})/2$
6. $c_{width(t)}$: Lateral "span" of c in time t
7. $W(t) = (\hat{c}_{diag} + c_{width(t)})/2$

I.4 Integrated Approach to the Safety of Automated Vehicles

As postulated in this series of edited books by the Society for Automotive Engineers (SAE) International, there are three main perspectives of safety when discussing the safety of AVs: functional safety, SOTIF, and multi-agent safety. At the highest level, the underlying concepts and understanding of safety are the same, *the absence or reduction of risk which can lead to accidents or harm to people*. However, at lower levels of abstraction, the approach of these safety perspectives when applied to autonomous vehicle safety is not the same, particularly when discussing multi-user safety. There are a large number of papers about the safety of AVs taking into account other agents in a dynamic fashion (i.e., other vehicles, pedestrians, moving obstacles, etc.). However, most of the literature address multiple-agent safety in an ad hoc manner and not integrated with the other safety categories of functional safety and SOTIF. Can all of these approaches or views be reconciled? Is there a unique framework for discussing the safety of AVs? In this section, we propose such a framework for integrating and unifying the discussion of safety at lower levels of abstraction when viewed from these three safety perspectives.

The unifying framework is based on dependability theory that includes fundamental principles of not only functional safety but reliability, availability, and security and which are based on cause and effects, that is, causal relationships [8]. According to functional safety precepts, the fundamental causes or sources of accidents are faults, errors, and failures which can lead to hazards, pose risks, and potentially lead to accidents. This chain of events is depicted in Figure 2. When untreated, these sources may represent a system that is highly unsafe. On the other hand, when properly treated and dealt with, the root causes of accidents can be treated or managed appropriately representing a highly safe system. We adopt the definitions of safety concepts drafted by well-known safety standards such as ISO 26262 [6]. We define *fault* as an "abnormal condition that can cause an element or an item to fail and it can be classified as permanent or intermittent." An *error* is a "discrepancy between a computed, observed or measured value or condition, and the true, specified or theoretically correct values." An error can arise as a result of unforeseen operating conditions or due to a fault. A *failure* is the "termination of the ability of an element or item to perform a function as required."

A *safety hazard* is defined as the "potential source of harm or accident caused by malfunctioning behavior of an item." A hazard is an abstract concept that needs the element to be protected in order to be realized or materialized and causes harm or accidents. When people or resources are exposed to hazards, the ensuing situation is called *risk*. There are many levels of risk, and to properly distinguish them, we need to know the *probability of exposure* to the hazard, the *severity* of the ensuing harm or accident, and the degree of avoidance of the specific harm or damage through the timely reactions of the drivers

FIGURE 2 Chain of safety events leading to an accident.

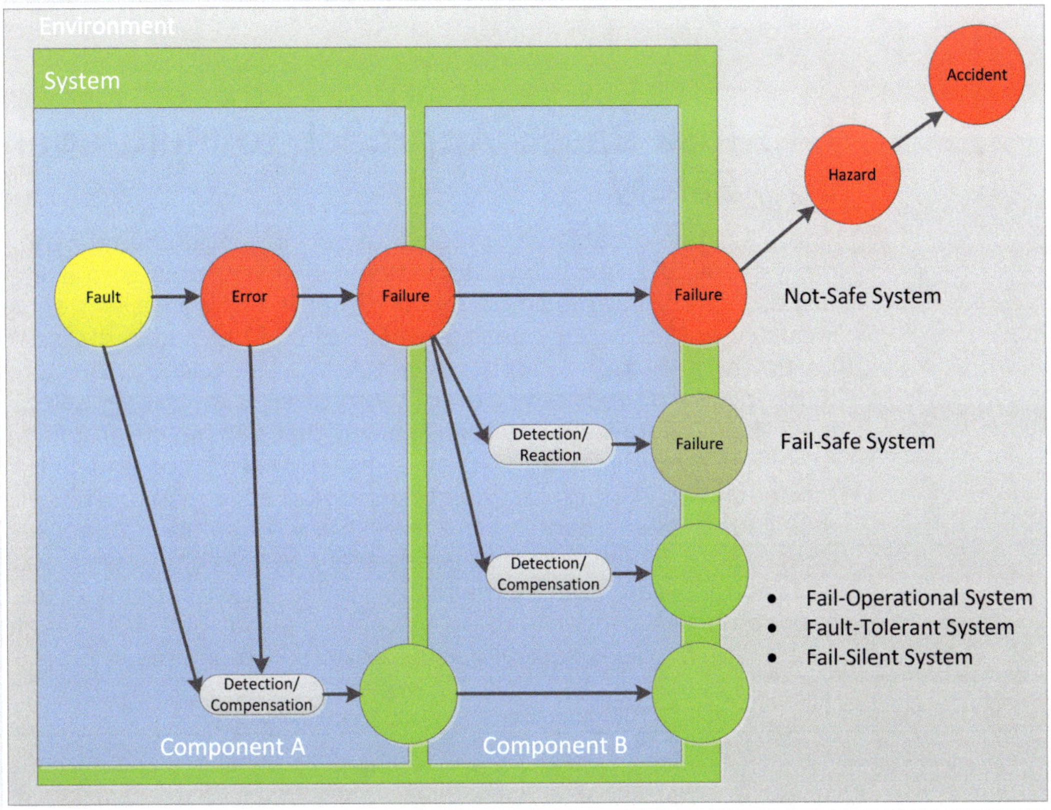

involved (called *controllability*). The following equation conveys this notion of various levels of risk:

$$R = S \times PE \times PC \tag{1}$$

where
R is the safety risk
PE is the probability of hazard occurrence (i.e., exposure)
S is the hazard severity
PC is the probability of controllability of the situation

Functional safety is defined as the "absence of unreasonable risk due to hazards caused by malfunctioning behavior of E/E systems." Thus functional safety deals only with malfunctions (mostly faults) of the very equipment that is brought in to enable automation, that is, electrical/electronic (E/E) systems. From this definition, it is clear the functional safety does not deal with hazards caused by temporal performance limitations (in the case of SOTIF) or hazards caused by other drivers (whether human or automated) or other agents as is the case in multi-user safety. As noted, when undetected or untreated, faults, errors, and failures lead to an unsafe system, as illustrated in the top portion of Figure 2. However, when detected, compensated, and treated in a reactive or proactive fashion, it is possible to design systems with various levels of safety ranging from a fail-safe system to a fail-operational system, to a fault-tolerant system, to a fail-silent system as depicted in the lower portions of Figure 2.

I.4.1 Risk-Based Approach to Multi-agent Safety

There exists a great deal of literature on multi-agent safety for AVs at the various SAE levels of automation. However, most of this literature discuss safety in an ad hoc manner that is not consistent across the industry and also not consistent with the other safety perspectives, particularly functional safety which follow a risk-based approach. In the following we discuss a model useful to discuss multi-user safety using a risk-based approach and thus in a way that can be integrated with the other safety perspectives, that is, functional safety and SOTIF. The model, shown in Figure 3, depicts a unifying approach for modeling the root causes of accidents of the three main AV safety categories: functional safety, SOTIF, and multi-user safety. We do not detail discussions on functional safety as this has been discussed extensively in the literature, rather we concentrate on discussing about SOTIF and multi-user safety. The model is basically an extension of the model of Figure 2 that

FIGURE 3 Unifying model for integrating SOTIF, multi-user safety, and functional safety.

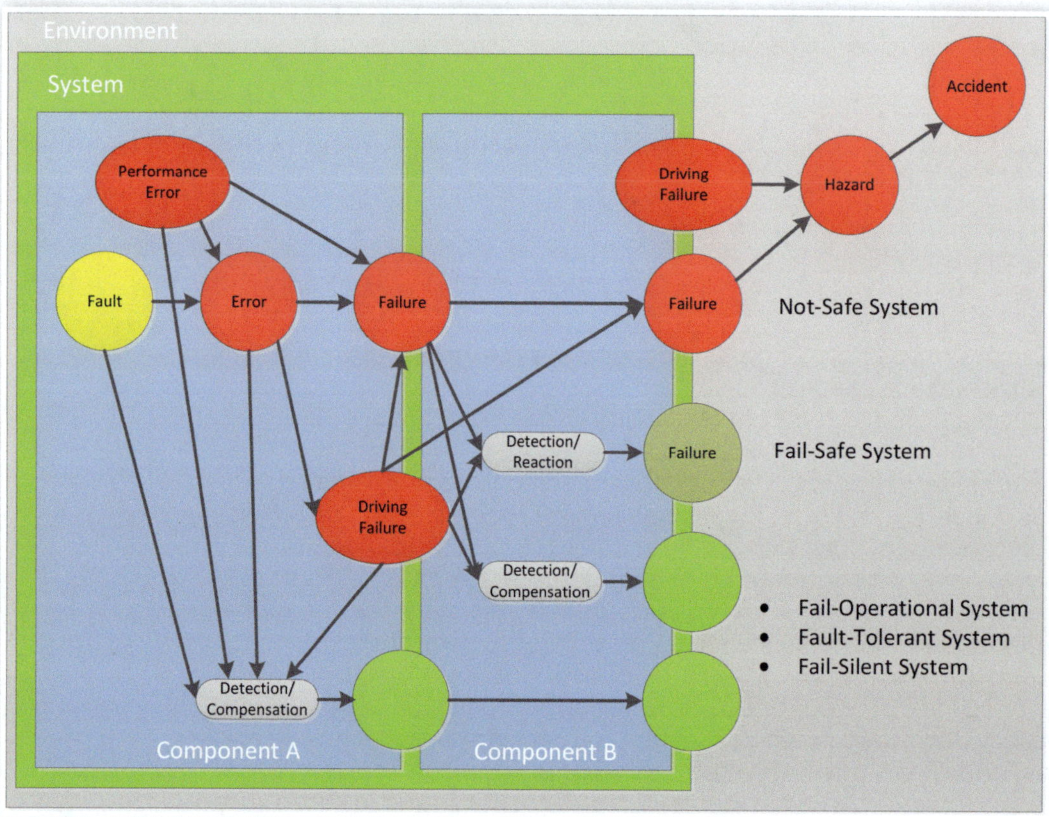

depicts the chain of events leading to accidents from the viewpoint of functional safety.

In terms of *SOTIF*, performance limitations of some functional units (e.g., the perception system of AVs) can lead to errors which in turn can lead to failures thus constituting hazards and eventually leading to accidents. A good example is performance limitations of a lidar sensor in bad weather, particularly when dealing with rain and fog, leading to false positives and false negatives. Thus, even perfectly functioning (i.e., no hardware and software faults) lidar units can be the source of errors which could lead to accidents. With this understanding of error and with additional definitions of some concepts, the entire theory and application of safety developed for functional safety can be used for modeling and discussing SOTIF. To reconcile SOTIF within the framework of functional safety, one can add another definition of *error* as the "discrepancy between a computed, observed or measured value or condition, and the true, specified or theoretically correct values strictly due to performance limitations of the measuring apparatus." Performance limitations are modeled in Figure 3 as *performance errors* which can lead to other errors and failures that can potentially cause accidents.

In terms *of multi-user safety*, the actions of other agents in the road or the actions of the computing system driver (whether a human or automated or whether a backup driver is involved) of the AV can be the source of failures or hazards that can lead to accidents as shown in Figure 3. This framework needs additional definitions and clarifications of the concepts of fault, error, failure, hazard, and risk so that they are consistent with those in functional safety. For the case of multi-user safety, another definition of *driving failure* can be added as follows: "termination of the ability of driver or the vehicle system to avoid hazardous situations." Likewise, another definition for *safety hazard* can be added as the "potential source of harm or accident caused by driving failures." As an example, consider the case of maintaining a *safe longitudinal distance* to a leading vehicle to avoid accidents. It is well known that there exists a minimum distance to a leading vehicle so that a collision with such vehicle can be avoided [20]. Such minimum distance is a function of the maximal braking and acceleration commands, the response time of the driver, the longitudinal velocities of the involved vehicles, and their lengths [20]. If such safe minimum distance is violated by either a human or automated driver, then the vehicle is exposed to risk that might lead to an accident. A violation on maintaining a safe distance to a leading vehicle can be due to an error in the calculation of safe distance (an error condition), driver distraction (a driving failure), a bug in the software (a driving failure), etc. The associated chain of events is shown in Figure 3.

Thus, by adopting a more general model for the underlying sources of accidents as depicted in Figure 3 and by adding additional definitions and concepts related to faults, errors, and failures, we can have a unified view of the safety of AVs that covers the three categories: functional safety, SOTIF, and multi-user safety. In summary, for the case of functional safety, hardware or software faults lead to errors and failures, resulting in safety hazards. For the case of SOTIF, performance degradation in the perception system can cause errors leading to failures involving safety hazards. Finally, for the case of multi-user safety, human or computing system-based drivers can cause failures leading to safety hazards.

I.4.2 HARA and Risk Reduction

The ISO 26262 standard details the phases of hazard analysis and risk assessment (HARA) and risk reduction which are critical to designing safe systems from the viewpoint of functional safety. Addressing HARA and risk reduction to designing safe systems from the viewpoint of SOTIF and multi-agent safety is not straightforward. This is so because in the case of functional safety this is done using Automotive Safety Integrity Level (ASIL) classifications for the various system components and designing specific safety mechanisms (i.e., risk reduction) according to the ASIL levels, and in the case of SOTIF and multi-user safety, the industry has not yet defined corresponding ASIL levels.

In the case of functional safety, ASIL is determined as a function of probability of exposure to the hazard, the severity of the potential harm or accident, and the controllability, as can be seen from Equation (1). This is done using 3 levels for severity, 4 levels for exposure, and 3 levels for controllability resulting in 3x4x3 = 64 levels for ASILs and classified as QM, ASIL A, ASIL B, ASIL C, and ASIL D with the last one being the most safety-critical. For traditional automotive applications, we have a wealth of knowledge and techniques on how to determine or estimate levels for severity, exposure, and controllability and thus how to estimate ASIL levels. No such set of knowledge and techniques exist for SOTIF and multi-agent safety, and we need much more research and work in this area in order to reconcile them with the wealth of knowledge that exists for functional safety.

I.5 RAND Safety Measurement Framework

In the wake of recent accidents, the industry is taking a fresh look at how to measure the safety levels of AVs. Recently, the RAND Corporation issued a report presenting a framework for measuring safety in AVs that could be used broadly by companies, policymakers, and the public [25]. The authors of this report discuss important issues such as how to define safety for AVs, how to measure safety for AVs, and how to communicate what is learned or understood about AVs. Since AVs currently have much less total on-road exposure compared with conventional, human-driven vehicles, the authors also consider options for proxy measurements—that is, factors that might be correlated with safety—and explore how safety measurements could be made in simulation and on closed courses.

The report identifies key concepts and paves the way for a better understanding of the kinds of measurements that might be made and communicated. It presents a structured way of thinking about how to measure safety at different AV lifecycle stages. The report highlights the kinds of information that could be presented in consistent ways in support of public understanding of AV safety, while acknowledging that the confidential nature of AV data limits the amount of data that are made public or shared between companies and with the government. The report addresses the safety of AVs from a public policy perspective and is based on observations on what companies that are key to the evolution of AVs have been doing to foster and evaluate the safety of those vehicles. As such, the report is intended for a broad public audience rather than design specialists. The authors considered how to define safety for AVs, how to measure safety for AVs, and how to communicate what is learned or understood about AVs. They also explored how safety measurements could be made in simulation and on closed courses.

Companies of all sorts and sizes ranging from OEMs to start-ups are in a race to design and deploy safe AVs. These companies are using a variety of methodologies, processes, standards, and design techniques to improve the level of safety of their AVs. However, the industry, policymaking community, and public can benefit from a higher level of discussion to understand and discuss the safety implications of AV technology. This is the viewpoint taken by the RAND Corporation report to present a framework to discuss how safety can be measured in a technology- and company-neutral way. Their work focuses on highly AVs while driving in typical driving scenarios. These vehicles are composed of features consistent with the widely accepted level 4 in the SAE taxonomy system. The vehicle can perform all driving functions within certain, prespecified conditions. It is noteworthy that the reports find that "the meaning of safety in regard to AVs is surprisingly unclear—no standard definition exists." The report defines safety as "the eliminating, minimizing, or managing of harm to the public" (which can include people, animals, and property, but the focus is on harm to people). In addition to the company involved in the design and testing of AVs, the public and the policymaking community have an important interest in comparing AV safety with the safety of conventional vehicles. As the reports point out, that comparison is difficult and complex because there are different limitations on the nature of data collected for AVs and for conventional vehicles.

To complicate matters, AVs have limitations on where and when they can operate; for example, level 4 AVs operate with full automation within parameters specified by the developers, which make up the operational design domain (ODD). An ODD can be defined in terms of geography, weather, lighting, roadway markings, previous experience on that specific roadway, and other factors. It is important that comparisons among AVs and between AVs and conventional vehicles must draw their data from the same ODD; otherwise results are likely biased.

Currently, we do not have AVs that belong to level 5 in the SAE taxonomy system with fully AVs only operating in small numbers in limited situations.

This state of affairs underscores the importance of understanding safety through the vehicle lifecycle stages of development, demonstration, and commercial deployment. The report considers a practice called *demonstration* (while acknowledging that there are other uses of that term) that is common in product development. For AVs, "it involves ensuring via targeted testing that what has been developed meets the ODD-specific environment's requirements and user needs, which include being safe." The report contends that because AV developers can be expected to do demonstration testing at intervals that mark milestones (other kinds of testing will take place at least regularly but perhaps continuously), those moments provide opportunities for communicating to the public about safety.

Of course there exits testing while the vehicle is in development prior to demonstration, and it involves challenging the system as a way to understand its remaining weaknesses. However the report does not address this kind of testing arguing that most results can reasonably be seen as internally useful to the developer. The report differentiates demonstration due to the following: (1) it provides an opportunity for demonstrating safety at or above a given level (allowing for some level of uncertainty and recognizing that there are limits to what can be demonstrated absent the accumulation of hundreds of millions or more miles driven); (2) its results could be made visible to external parties (something that can differentiate it from routine validation); and (3) it could be undertaken using a protocol that is (semi- or completely) uniform across AVs (building on common attention to industry standards). Such a protocol could involve sets of AV activities in a prespecified sequence and environment that developers call scenarios.

The RAND report is high level taking the concept and application of safety from the broadest point of view [25]. Although such a view is important to discuss issues with multiple stakeholders, it is not detailed enough to discuss more specific issues, particularly design issues. The report presents a framework for measuring safety in AVs that could be used broadly by companies, policymakers, and the public. Such framework extends general notions, ideas, and practices of measuring safety in conventional vehicles where a human driver is in command to measuring safety in AVs where a computing system is in command. The framework includes measures and metrics that are classified according to the lifecycle of a typical AV and the place (i.e., where) the AV is being evaluated. The lifecycle stages are development, demonstration, and deployment. As the place of evaluation, the reports consider with a computer (simulation), in a closed course, and in public roads.

The developed framework is shown in Figure 4 and depicts safety measures in three lifecycle stages (development, demonstration, and deployment) and three places of testing (computer (i.e., simulation), closed course, and public roads). Measures must be valid, feasible, reliable, and non-manipulatable. Two types of safety measures are included: *leading* and *lagging*. The former reflect performance, activity,

FIGURE 4 Safety measures for the safety measurement of automated vehicles developed by the RAND Corporation.

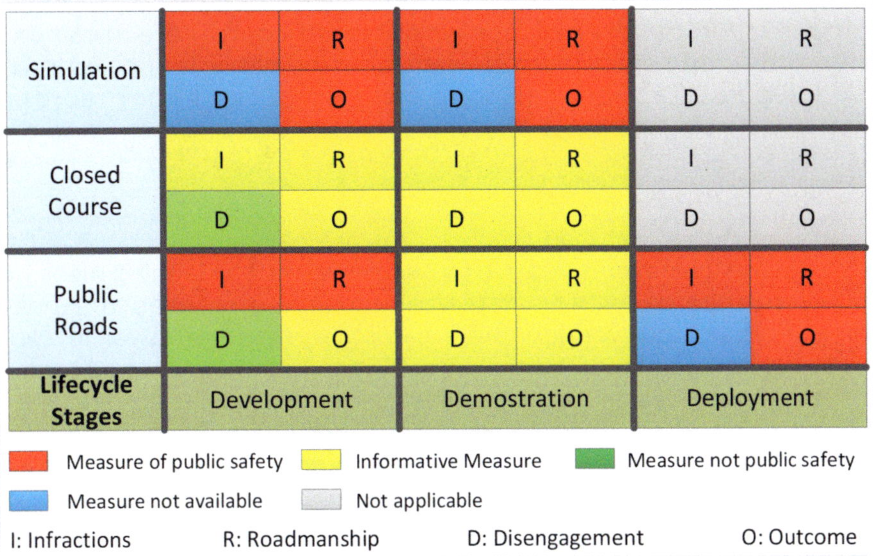

and prevention in the precrash phase, and the latter are observations of safety outcomes involving harm. Examples of leading measures can be proxy measures of driving behaviors correlated to safety outcomes. Leading measures could include *infractions* (failures to follow traffic rules), an integrated measure of driving abilities that are called *roadmanship*, and *disengagements* (occasions when a person has to take over the control of the vehicle from the automated system). Lagging measures are *outcome* measures, including crashes, injuries, deaths, and cost of property or vehicle repairs or of medical expenses (for both short and long terms). A third type of measurement, around standards, processes, procedures, and design, is not included because it can be applied at lower levels of vehicle composition (i.e., not for complete vehicles) that evolve constantly and because these measures relate indirectly to safety through function. One example of a safety measure that can be used during design is the number of expected catastrophic failures per year of operations (e.g., 500 hours) [24].

Although it is conceivable that a measure exists for each combination of lifecycle stage and place of testing, some of these do not make sense (i.e., not applicable), for example, testing deployed vehicles only makes sense on public roads and not in a closed course or in simulation. The interpretation and utility of using a specific measure differs for a given lifecycle stage and place of testing. For example, a measure of infractions on public roads during development might indicate the vehicle's ability to act without legal violations within an environment where the developer has little control beyond ODD specifications. A measure of infractions in closed-course testing during demonstration might indicate the vehicle's ability to operate legally within predesignated maneuvers or scenarios, gauging readiness for public deployment.

RAND's proposed framework focuses on measures that reflect public safety—that is, the risks to which the public is exposed [25]. However, measures for which the public is not at risk (during development and demonstration using simulation or at closed course) are obtained in artificial environments where, by definition, the public is not exposed to risk. Measures for these cases are included because they represent gateways for development on public roads and deployment, and their measures are determinative of safety, even if they are not themselves situations in which public exposure occurs.

The report notes that leading measures are particularly important for AVs because their events happen more frequently than lagging measures. This frequency allows for statistically meaningful, exposure-based comparisons to conventional vehicles to be made at lower levels of exposure (i.e., accumulated mileage). Consequently, an understanding of the AV's safety can be considered earlier in the development and deployment stages. Measures unlikely to accumulate sufficient occurrences in a given stage and setting to statistically meaningful measures, and whose events could instead serve as a basis for case studies, are noted in yellow in Figure 4. The report also emphasizes that clearer communication about safety between the industry and the public will be critical for public acceptance of AVs calling for more consistency in the communication around AV safety from industry as this will result in a message that is more cohesive and comprehensible. Clarity is noted as needed on how to think about AV safety (both in absolute terms and in terms that are relative to conventional vehicles), how to measure safety when the AV system continually updates, and how to maximize safety in the face of unknowns.

In addition to the framework, the report offers some recommendations; the most noteworthy are:

- *Given the potential for broader learning across industry and government, a protocol for information-sharing should be encouraged. It would have to precisely incorporate measures, format, context, frequency, governance, data security, and other factors.*
- *A taxonomy for common use that facilitates understanding of and communication about operational design domains is needed. A common approach to specifying where, when, and under what circumstances an AV can operate would enable, in particular, inter- and intraorganizational communication and communication with consumers and regulators. It would also facilitate tracking for a given AV of its progress through development and into deployment. Minimal-risk conditions should also be included. Such taxonomies are currently under consideration by industry groups.*
- *Research is needed around how to measure and communicate AV system safety in an environment wherein the system evolves through frequent updates. AV safety measures must balance reflecting the current system's safety level with recent (and perhaps non-recent) safety records.*

I.6 The Papers in This Collection

There are actually a large number of papers in the area of multi-agent safety from SAE and other sources. The following is a representative list of ten SAE papers on the subject.

1. Helene G. Moorman, Andrea Niles, Caroline Crump, Benjamin Lester, Laurene Milan, Christy Cloninger, David Cades, Douglas Young, and Audra Krake, "Lane-Keeping Behavior and Cognitive Load with Use of Lane Departure Warning," SAE Technical Paper 2017-01-1407, 2017, doi:10.4271/2017-01-1407.
2. Lin Liu, Xichan Zhu, and Zhixiong Ma, "Study on Test Scenarios of Environment Perception System under Rear-End Collision Risk," SAE Technical Paper 2018-01-1079, 2018, doi:10.4271/2018-01-1079.
3. Dongfang Dang, "Motion Planning of Vehicle Obstacle Avoidance in Complex Traffic Scenarios," SAE Technical Paper 2017-01-1958, 2017, doi:10.4271/2017-01-1958.
4. John Shutko, Benjamin Osafo-Yeboah, Chris Rockwell, and Mark Palmer, "Driver Behavior While Operating Partially Automated Systems: Tesla Autopilot Case Study," SAE Technical Paper 2018-01-0497, 2018, doi:10.4271/2018-01-0497.
5. Nazan Aksan, Lauren Sager, Sarah Hacker, Robert Marini, Jeffrey Dawson, Steven Anderson, and Matthew Rizzo, "Forward Collision Warning: Clues to Optimal Timing of Advisory Warnings," SAE Int. J. Trans. Safety 4(1):2016, doi:10.4271/2016-01-1439.
6. Madeleine Gibson, John Lee, Vindhya Venkatraman, Morgan Price, Jeffrey Lewis, Olivia Montgomery, Bilge Mutlu, Joshua Domeyer, and James Foley, "Situation Awareness, Scenarios, and Secondary Tasks: Measuring Driver Performance and Safety Margins in Highly Automated Vehicles," SAE Int. J. Passeng. Cars - Electron. Electr. Syst. 9(1):2016, doi:10.4271/2016-01-0145.
7. Frank Keck, Marco Alt, Arne Vater, and Joseph Wessner, "Frontal Collisions - What are the Limitations of Future Forward-Looking Safety Systems?," SAE Technical Paper 2017-26-0013, 2017, doi:10.4271/2017-26-0013.
8. Adit Joshi, "Hardware-in-the-Loop (HIL) Implementation and Validation of SAE Level 2 Autonomous Vehicle with Subsystem Fault Tolerant Fallback Performance for Takeover Scenarios," SAE Technical Paper 2017-01-1994, 2017, doi:10.4271/2017- 01-1994.
9. Keyu Qian, Gangfeng Tan, Renjie Zhou, Binyu Mei, and Wanyang Xia, "Vehicle-GIS Assistant Driving System for Real-time Safety Speed Warning on Mountain Roads," SAE Technical Paper 2017-01-1400, 2017, doi:10.4271/2017-01-1400.
10. Ting Jun Li, Minghao Shen, and Hongyu Zheng, "Overtaking or Merging? Eco-Routing Decision and Speed Trajectory with Full Terrain Information," SAE Technical Paper 2018-01-0038, 2018, doi:10.4271/2018-01-0038.

In the following, we summarize the above papers in the context of the characterization of the safety of self-driving vehicles. After this introduction, the actual papers follow.

I.6.1 Discussion of the Multi-agent Safety Papers

1. Helene G. Moorman, Andrea Niles, Caroline Crump, Benjamin Lester, Laurene Milan, Christy Cloninger, David Cades, Douglas Young, and Audra Krake, "Lane-Keeping Behavior and Cognitive Load with Use of Lane Departure Warning."

   Observations from the authors suggest that a more complete transition to fully autonomous driving systems, rather than to partially or highly autonomous systems (i.e., going from level 2 directly to level 4), may ultimately be more acceptable to drivers. This paper maintains that drivers have generally reported satisfaction with ADAS with the exception of lane departure warning (LDW) systems, which are often rated poorly or even deactivated by drivers. The authors argue that one potential contributor to this negative response may be an increase in the cognitive load associated with lane-keeping when LDW is in use. The paper examined the relationship between LDW, lane-keeping behavior, and concurrent cognitive load, as measured by performance on a secondary task. Participants drove a vehicle equipped with LDW in a demarcated lane on a closed-course test track with and without the LDW system in use over multiple sessions. The findings suggest that LDW alerts are associated with impairments in performance on a secondary cognitive task, possibly because they increase the cognitive load associated with maintaining lateral vehicle control. This may partially explain why drivers have demonstrated lower levels of acceptance for LDW when compared to other types of ADAS.

2. Lin Liu, Xichan Zhu, and Zhixiong Ma, "Study on Test Scenarios of Environment Perception System under Rear-End Collision Risk."

   In this paper, the authors calculate the probability distribution of environmental objects correlated with risk-level scenarios for rear-end collisions. This may pave the way to perform HARA for multi-agent safety in an integrated way of that used in functional safety. The authors define an environment perception element (EPE) as any object that must be perceived during driving including weather and light conditions, road features, road markings, traffic signs, traffic lights, other vehicles, pedal cyclists, and pedestrians. Driving behaviors collected in Shanghai, China, were classified, and rear-end collision risk scenarios were obtained and described using EPEs.

3. Dongfang Dang, "Motion Planning of Vehicle Obstacle Avoidance in Complex Traffic Scenarios."

   This paper provides a good example on how to design a risk reduction mechanism, more specifically how to design a vehicle control system that minimizes risk. The results show that the control system can safely handle a variety of complex traffic scenarios, such as a cut-in situation, while meeting traffic regulations and vehicle dynamic constraints. The authors present a general driving risk model that includes (i) collision risk, that is, the collision risk between the subject vehicle (SV) and other surrounding vehicles, pedestrians, buildings, etc., and (ii) non-collision risk, such as violation of traffic regulations, the deviation from the intention of driver, etc. An index similar to the exponential function is defined to calculate the collision risk value, which is composed of a real-time calculation of collision detection and time to collision (TTC). The index of non-collision risk is measured by the deviation from the desired states including regulations, driving manners and intentions, etc.

4. John Shutko, Benjamin Osafo-Yeboah, Chris Rockwell, and Mark Palmer, "Driver Behavior While Operating Partially Automated Systems: Tesla Autopilot Case Study."

   This paper details a study of driver behavior that helps understanding driving risks of AVs at SAE levels 1, 2, and 3. One of the main challenges facing developers of these systems is how to ensure that drivers understand their role and stay alert as the systems require. In order to understand how customers interact with a partially AV, a study was conducted to observe people driving their own Tesla vehicles while autopilot was engaged. Results from the study show that (i) participants' eye off-road glance behavior while operating a partially AV was somewhat similar to eye off-road glance behavior in radio tuning task while driving manually, though substantially more variable, (ii) eyes on-/off-road glance behavior had no relation to whether drivers kept their hands on the steering wheel or not, and (iii) drivers exhibit a bimodal behavior when operating a partially AV: those who mostly kept their hands on steering wheel while autopilot was active (i.e., *active drivers*) versus those who mostly kept their hands off the steering wheel while autopilot was active (i.e., *supervisors*).

5. Nazan Aksan, Lauren Sager, Sarah Hacker, Robert Marini, Jeffrey Dawson, Steven Anderson, and Matthew Rizzo, "Forward Collision Warning: Clues to Optimal Timing of Advisory Warnings."

   This paper provides a case study that will help performing HARA particularly involving risk tolerance behavior of different age groups and whether risk tolerance varies as a function of cognitive functioning. The authors examined the effectiveness of a heads-up FCW system in 39 younger to middle-aged drivers (25-50, mean = 35 years) and 37 older drivers (66-87, mean = 77 years). The warnings were implemented in a fixed based, immersive, 180-degree forward field of view simulator. The FCW included a visual advisory component consisting of a red horizontal bar which flashed in the center screen of the simulator that was triggered at TTC of 4 seconds. The findings suggest optimal timing for advisory alerts for forward events may need to be larger than TTC=4 especially for older drivers.

6. Madeleine Gibson, John Lee, Vindhya Venkatraman, Morgan Price, Jeffrey Lewis, Olivia Montgomery, Bilge Mutlu, Joshua Domeyer, and James Foley, "Situation Awareness, Scenarios, and Secondary Tasks: Measuring Driver Performance and Safety Margins in Highly Automated Vehicles."

   This paper provides further insight into performing HARA in multi-agent safety. The authors point out that risk analysis typically identifies failure modes associated with the mechanical, electronic, and software elements of systems and works to assure those risks remain below an acceptable level. However, such risk estimates assume that drivers provide an additional safety margin, compensating for failures of the technology and other sources of unanticipated variability. The authors argue that this assumption is not always justified and just as technology has failure modes, so do drivers and these failure modes should be assessed as part of a simulator-based evaluation. Implicitly or explicitly vehicle automation encourages drivers to disengage from driving and engage in other activities that may result in safety risks. Thus secondary tasks play an important role in both creating representative situations for automation use and misuse and providing embedded measures of driver engagement. Latent hazards—hazards that exist in the road environment and merit driver attention but do not materialize to require a driver response—have been used with great success for understanding the vulnerability of novice drivers. Latent hazards might provide a similarly useful index of driver attention to the road during periods where the automation is vulnerable to failure. With highly AVs, latent hazards include potential roadway threats that might not be sensed by the automation and would require driver attention. This paper describes driving simulator scenarios used to operationalize automation-relevant latent hazards, secondary tasks tuned to index driver disengagement from the driving task, and measures that reflect safety margins rather than driving performance, such as drivers' trust, situation awareness, and expected time to transition to manual control.

7. Frank Keck, Marco Alt, Arne Vater, and Joseph Wessner, "Frontal Collisions-What are the Limitations of Future Forward-Looking Safety Systems?"

   This paper uses a stochastic formulation of frontal collisions using measures such as probability density of TTC and probability of no crash as a function of distance to target. Using this mathematical framework in one and two dimensions, the authors consider several critical scenarios and address how should a system be designed to be able to handle these situations including limitations which cannot be overcome. These limitations are mainly incomplete information that the ego vehicle has about other vehicles and their intentions. The shortages of sensors, system latencies, the stochastic nature of traffic scenarios, the unknown behavior of the driver, and the high dynamics when considering oncoming traffic all contribute to this lack of information. The authors suggest that other measures such as sharing information using Car2Car communications or a centralized path planning for autonomous cars will help reduce collisions.

8. Adit Joshi, "Hardware-in-the-Loop (HIL) Implementation and Validation of SAE Level 2 Autonomous Vehicle with Subsystem Fault Tolerant Fallback Performance for Takeover Scenarios."

   This research presents system modes in which longitudinal control only and lateral control only are engaged as fallback states to the full autonomous system being faulted for lateral control and longitudinal control failures, respectively, using a hardware simulation platform consisting of a dSPACE hardware-in-the-loop (HIL) simulator and CarSim RT. The research was also used to provide a real-time powertrain-based HIL implementation and validation of an SAE level 2 autonomous vehicle. The fallback performance was evaluated and validated by introducing subsystem faults when the AV was in autonomous mode. This would force the automated system into the fallback states.

The vehicle controller includes an acceleration-based adaptive cruise control (ACC) combined with a path-following lateral control along with supervisory control for system mode transitions due to system deactivations and faults. Results show that the automated lateral control continues to operate by enabling trajectory tracking even though there was a fault in the automated longitudinal control causing loss of the latter subsystem. Analysis in the paper showed the importance of the lateral and longitudinal controls being independent and thereby increasing reliability in case of failure of one subsystem.

9. Keyu Qian, Gangfeng Tan, Renjie Zhou, Binyu Mei, and Wanyang Xia, "Vehicle-GIS Assistant Driving System for Real-time Safety Speed Warning on Mountain Roads."

   This paper uses vehicle dynamics together with mapping and localization to improve the driving safety of vehicles when driving on mountain roads, particularly during downhill situations. Simulation results show that a prewarning scheme works well for downhill roads, especially with length from 1 km to 1.5 km and with slope gradient ranging between 3% and 6%. This kind of study is important because downhill mountain roads are the accident-prone sections due to their complexity and variety. First, the characteristics of the driving conditions under the change of various slope lengths and gradients are studied based on models of both engine braking and wheel braking. The real-time information and characteristics of the vehicle on the mountainous area, such as the weight, the speed, the gear, and the engine RPM, and the vehicle braking torque characteristics on the mountainous area are then analyzed. In addition, the electronic map and GPS positioning are utilized to calculate the road gradient information. Finally, geographic information, road traffic speed limits, and the brake efficiency will be combined to obtain a suggested safe speed at the top of the downhill roads and give a safety speed prewarning to the driver.

10. Ting Jun Li, Minghao Shen, and Hongyu Zheng, "Overtaking or Merging? Eco-Routing Decision and Speed Trajectory with Full Terrain Information."

    This article studies optimal speed trajectories and longitudinal control with optimized energy efficiency for an autonomous vehicle in a platoon configuration taking into account the safety of the overtake or merge actions. A forward discrete dynamic programming (DDP) algorithm with distance as the discretization basis is used to derive speed trajectories in the trade-off between air drag reduction and energy saved by utilizing the road slope information. The problem is decomposed into decisions whether to overtake or to merge into the nearest platoon with the assumption of sufficient distance among platoons due to safety considerations. During the process, speed choices and cost function reflect interactions among the controlled vehicle, the platoons, and the road. Simulations confirm that the energy consumed can be reduced significantly when tracking optimal trajectories compared to driving by the existing model predictive control (MPC) tracking strategy and linear quadratic regulator (LQR). Simulations indicate the potential of the proposed method for achieving energy efficiency in further studies that include more sophisticated conditions. Sensitivity analysis has shown that the energy saving is not sensitive to the initial speed of the ego car when it is relatively low.

I.7 Conclusion

Multi-agent safety is an important safety category that is perhaps the most visible to the general public and complements the categories of SOTIF and functional safety. Multi-agent safety involves addressing hazards due to actions of agents such as other vehicles, pedestrians, bicyclists, and in general any object that is external to the vehicle. Much of the literature on multi-agent safety is not addressed using a risk-based approach as it is the case with functional safety. In this chapter, we have characterized multi-agent safety and summarized the concept of RSS. In addition, an integrated framework to safety is presented which unifies the three safety categories (functional safety, SOTIF, and multi-user safety) using a risk-based approach. In addition, we summarized a recent report from the RAND Corporation on measuring the safety levels of AVs.

The proposed RAND framework does not consider measures related to standards, processes, procedures, and design because these are "evolving constantly and might be measured at lower levels of vehicle composition (e.g., at the subcomponent level)—and because measures of adherence relate indirectly to safety through function." This category of measure shows design approaches through which it is possible to track how safety is built in. Thus the proposed framework helps little at the design and development stages of AVs. Nevertheless, the report acknowledges that adherence to industry standards is "both industry best practice and associated with reducing potential liability for companies." As such, they are at a high level and thus it is difficult to determine how the safety measures can be traced back to hazards related to functional safety, SOTIF, or multi-agent safety. Nevertheless, due to the nature of these metrics, it is clear that they mostly deal with multi-agent safety. Thus safety measures applicable for lower levels of design and vehicle composition are highly needed, particularly safety measures in the categories of functional safety, SOTIF, and multi-agent safety.

The RAND study safety measures provide context for using a measure such as the number (in millions) of miles driven which needs to be qualified in terms of the number of incurred infractions, roadmanships, disengagements, and accidents. Furthermore, the above needs to be classified into a lifecycle stage (development, demonstration, or deployment), place of test (simulation, closed course, or public road), and the degree of complexity of the test.

References

1. Thrun, S. et al., "Stanley: The Robot that Won the DARPA Grand Challenge," *Journal of Field Robotics* 23, no. 9 (2006): 661–692.
2. Urmson, C. et al., "Autonomous Driving in Urban Environments: Boss and the Urban Challenge," *Journal of Field Robotics* 25, no. 8 (2008): 425–466.
3. Cheng, H., *Autonomous Intelligent Vehicles: Theory, Algorithms, and Implementation* (London, Springer, 2011).
4. Wang, F.-Y. et al., "IVS O5: New Developments and Research Trends for Intelligent Vehicles," *IEEE Intelligent Systems* 20, no. 4 (2005): 10–14.
5. Cheng, H. et al., "Interactive Road Situation Analysis for Driver Assistance and Safety Warning Systems: Framework and Algorithms," *IEEE Transactions on Intelligent Transportation Systems* 8, no. 1 (2007): 157–166.
6. International Organization for Standardization, "Road Vehicles—Functional Safety," ISO Standard 26262, 2011.
7. Koopman, P. and Wagner, M., "Autonomous Vehicle Safety: An Interdisciplinary Challenge," *IEEE Intelligent Transportation Systems Magazine* 9, no. 1 (2017): 90–96.
8. Avizienis, A. et al., "Basic Concepts and Taxonomy of Dependable and Secure Computing," *IEEE Transactions on Dependable and Secure Computing* 1, no. 1 (2004): 11–33.
9. Leveson, N.G., *Safeware: System Safety and Computers* (Addison-Wesley, 1995).
10. Wendorff, W., "Quantitative SOTIF Analysis for Highly Automated Driving Systems," *Safetronic.2017, Conference Proceedings*, Stuttgart, Germany, 2017.
11. Thomas, J., Sgueglia, J., Suo, D., Leveson, N. et al., "An Integrated Approach to Requirements Development and Hazard Analysis," SAE Technical Paper 2015-01-0274, 2015, doi:10.4271/2015-01-0274.
12. Leveson, N.G., *Engineering a Safer World : Systems Thinking Applied to Safety* (MIT Press, 2012).
13. Young, W. and Leveson, N.G., "An Integrated Approach to Safety and Security Based on Systems Theory," *Communications of the Association for Computing Machinery (ACM)* 57, no. 2 (2014):31–35.
14. Abdulkhaleq, A., Wagner, S., and Leveson, N., "A Comprehensive Safety Engineering Approach for Software-Intensive Systems Based on STPA," *3rd European STAMP Workshop, Conference Proceedings*, Amsterdam, The Netherlands, 2015.
15. Abdulkhaleq, A. et al., "A Systematic Approach Based on STPA for Developing a Dependable Architecture for Fully Automated Driving Vehicles," *4th European STAMP Workshop, Conference Proceedings*, Zurich, Switzerland, 2017.

16. Kopetz, H., *Real-Time Systems: Design Principles for Distributed Embedded Applications* (Kluwer Academic Publishers, 1997).

17. Koopman, P. and Wagner, M., "Toward a Framework for Highly Automated Vehicle Safety Validation," SAE Technical Paper 2018-01-1071, 2018, doi:10.4271/2018-01-1071.

18. International Electrotechnical Commission, "Functional Safety of Electrical/Electronic/Programmable Electronic Safety-Related Systems," IEC Standard 61508, 2010.

19. International Electrotechnical Commission, "Functional Safety—Safety Instrumented Systems for the Process Industry Sector," IEC Standard 61511, 2018.

20. Shalev-Shwartz, S., Shammah, S., and Shashua, A., "On a Formal Model of Safe and Scalable Self-Driving Cars," *Computing Research Repository (CoRR)*, vol. arXiv:1708.06374 [cs.RO] (2017). [Online]. http://arxiv.org/abs/1708.06374.

21. Kim, K. et al., "Design of Integrated Risk Management-Based Dynamic Driving Control of Automated Vehicles," *IEEE Intelligent Transportation Systems Magazine* 9, no. 1 (2017): 57–73.

22. Feth, P., Adler, R., Fukuda, T., Ishigooka, T. et al., *Multi-Aspect Safety Engineering for Highly Automated Driving Looking Beyond Functional Safety and Established Standards and Methodologies*, Gallina, B. et al. (Eds.): *SAFECOMP 2018, LNCS 11093*, 2018, 59–72.

23. Pimentel, J. and Bastiaan, J., "Characterizing the Safety of Self-Driving Vehicles: A Fault Containment Protocol for Functionality Involving Vehicle Detection," *2018 IEEE International Conference on Vehicular Electronics and Safety (ICVES)*, Madrid, Spain, September 12-14, 2018.

24. Pimentel, J., Bastiaan, J., and Zadeh, M., "Numerical Evaluation of the Safety of Self-Driving Vehicles: Functionality Involving Vehicle Detection," *2018 IEEE International Conference on Vehicular Electronics and Safety (ICVES)*, Madrid, Spain, September 12-14, 2018.

25. Fraade-Blanar, L., Blumenthal, M.S., M. Anderson, J., and Kalra, N., *Measuring Automated Vehicle Safety: Forging a Framework* (Santa Monica, CA: RAND Corporation, 2018), https://www.rand.org/pubs/research_reports/RR2662.html, Also available in print form.

CHAPTER 1

Lane-Keeping Behavior and Cognitive Load with Use of Lane Departure Warning

Helene G. Moorman, Andrea Niles, Caroline Crump, Benjamin Lester, Laurene Milan, Christy Cloninger, David Cades, and Douglas Young
Exponent Inc.

Audra Krake
California State University Long Beach

Lane Departure Warning (LDW) systems, along with other types of Advanced Driver Assistance Systems (ADAS), are becoming more common in passenger vehicles, with the general aim of improving driver safety through automation of various aspects of the driving task. Drivers have generally reported satisfaction with ADAS with the exception of LDW systems, which are often rated poorly or even deactivated by drivers. One potential contributor to this negative response may be an increase in the cognitive load associated with lane-keeping when LDW is in use. The present study sought to examine the relationship between LDW, lane-keeping behavior, and concurrent cognitive load, as measured by performance on a secondary task. Participants drove a vehicle equipped with LDW in a demarcated lane on a closed-course test track with and without the LDW system in use over multiple sessions. On a subset of laps, participants were required to perform a secondary task requiring mental arithmetic while driving. The frequency and duration of lane line crossings and the participants' accuracy on the secondary arithmetic task were recorded for each lap in two sessions. LDW successfully enhanced lane-keeping behavior in the earlier session, but this effect was eliminated in the later session. Secondary task performance showed no effect of LDW on the earlier

CITATION: Moorman, H., Niles, A., Crump, C., Lester, B. et al., "Lane-Keeping Behavior and Cognitive Load with Use of Lane Departure Warning," SAE Technical Paper 2017-01-1407, 2017, doi:10.4271/2017-01-1407.

session, but significantly declined when LDW was in use during the second session. Implications of these preliminary results are discussed with regard to overall driver performance with LDW and other ADAS, as well as driver acceptance of more fully autonomous vehicle systems.

Introduction

Lane Departure Warning (LDW) systems, along with other types of Advanced Driver Assistance Systems (ADAS), are becoming more common in passenger vehicles [1, 2, 3]. An important goal for these technologies is to improve driver safety through automation of various aspects of the driving task. In line with this goal, LDW systems provide alerts (typically auditory beeps) to drivers when lane-line crossings are detected. Lane departure is a factor in a large number of traffic accidents-in 2014, approximately 17% of all crashes and 40% of fatal crashes occurred off the roadway [4]. LDW may potentially reduce the number of these types of incidents by making drivers aware of departures early enough for them to correct their steering and avoid a crash.

Various studies have reported a number of positive effects associated with LDW, including reductions in crashes [5], increases in turn signal use [6], improved lane-keeping [6, 7, 8], and reduced reaction times in drowsy drivers [9, 10]. In evaluating the overall benefit of such systems, however, driver acceptance is an important factor that must be considered. Compared to other ADAS, drivers more often report dissatisfaction or annoyance with LDW systems, and have been observed to deactivate the systems during normal use, rendering any potential safety benefits moot [11, 12, 13].

One potential contributor to negative driver response to LDW may be that it induces an increase in the cognitive load associated with lane-keeping. For experienced drivers, the driving task has been characterized as a type of highly "automatic" (fast and effortless, requiring few attentional resources) behavior punctuated by instances of more "controlled" (slow and demanding of attention) behavior when novel, unexpected, or uncertain situations arise [14]. LDW may be perceived by drivers as increasing the difficulty of the driving task by requiring more frequent interruptions to the automated driving process in which attention and response to LDW alerts are required. Some support for this notion comes from a study by Dinges et al. [15], in which a suite of fatigue management technologies which provided feedback to drivers, including LDW, were tested for their effects on driver performance and alertness. The authors found that driver drowsiness and lane tracking variability were decreased as a result of the intervention, but suggested that there may also have been an associated "cost" in the form of added effort required to respond to the feedback information.

Several previous studies have explored general questions of cognitive load in the context of ADAS. For example, Young and Stanton [16] measured drivers' accuracy on a concurrent cognitive task under various conditions of automated longitudinal (Adaptive Cruise Control) and lateral (Active Steering) vehicle control. They observed better performance (and thus lower cognitive load) during the automated conditions when compared to the non-automated, "manual" condition. Merat et al. [17] examined cognitive task performance under similar automated driving conditions involving Adaptive Cruise Control and a Lane Keeping Assist System which maintained the vehicle's lateral position in the center of the lane. In that study, the authors observed higher performance (by one measure) on the cognitive task during the automated driving condition vs. the manual condition when the driving requirements were low, but the effect was reversed when participants were required to respond to a "critical incident" (a lane obstruction necessitating a lane change) that was introduced to the driving task. Additional studies have presented similar evidence of improved secondary-task performance during use of Adaptive Cruise Control [18, 19]. Taken together, studies suggest that active systems (e.g., Adaptive Cruise Control and Active Steering or Lane Keeping Assistance) reduce the driver's cognitive load, allowing the driver to maintain the same level of driving performance (at least in the absence of abnormal or challenging driving conditions) using fewer attentional resources than under traditional "manual" driving conditions.

None of the above studies, however, examine LDW specifically. Whereas Adaptive Cruise Control and Active Steering function by detecting an undesirable condition (deviation in longitudinal and lateral position, respectively) and then automatically correcting for it without need of input from the driver, LDW performs only the first part of this process; that is, the system detects the undesirable condition (lateral deviation), but does not correct for it. Instead, feedback is provided to the driver, who must then interpret and understand the signal, decide on a response, and respond with an appropriate steering input. The accompanying attentional requirements and effects on cognitive load may differ from (or even contradict) those observed for more fully automated ADAS. The degree to which LDW increases cognitive requirements associated with driving therefore remains an open question.

The present study sought to explore this question by examining the relationship between LDW, lane-keeping behavior, and cognitive load as measured by performance on a mental math task in a preliminary dataset. Drivers performed the mental math task while driving with and without LDW over several experimental sessions. Performance on the math task was compared when LDW was active and when it was not. We also examined the effects of LDW on lane-keeping behavior independent of the cognitive task. We hypothesized that LDW would improve lane-keeping, but impair performance on the cognitive task. We propose that impaired performance on the cognitive task during LDW activation would reflect a shifting of cognitive resources to compensate for an increased cognitive load associated with responding to LDW alerts.

Methods

Participants

Twenty adults (15 female; median age 46.5 years; age range: 32-57) recruited from the Phoenix, Arizona area participated in this study. Although attempts were made to achieve participant gender balance, recruitment and scheduling constraints led to the use of a convenience sample with more females than males. All participants had a valid driver's license and had normal or corrected-to-normal vision. Participants were screened to ensure that they did not currently drive a vehicle equipped with ADAS technologies including ACC, LDW, FCWM, and similar technologies such as automatic braking and driver alert/fatigue-sensing systems. Each participant performed 3 driving sessions (Visits 1, 2, and 3), separated by approximately a week.

Materials

The experiment was conducted on a two-mile oval track. A single lane was painted on the track for testing. On each of the two straightaways on the test track, the lane shifted at the halfway point and transitioned either from wide to narrow or narrow to wide, to

FIGURE 1 Schematic of the test track and driving lane (not drawn to scale).

increase the difficulty of the lane-keeping task (see Figure 1). The narrow sections of the lane markings were approximately seven feet in width and the wide sections were approximately 11 feet in width.

The test vehicle was a 2014 luxury sedan equipped with LDW. The LDW system consisted of both auditory (series of quickly-repeating tones) and visual (digital display of the vehicle leaving a lane on the instrument cluster) warnings when sensors detected the vehicle was moving outside of the lane markings without use of a turn signal. The volume of the auditory warnings began low and increased as time outside the lane increased. There were no instances where experimenters observed participants either failing to hear/notice the warnings or exhibiting a startle response to their occurrence. LDW was engaged or disengaged by the experimenter on the appropriate laps, using the vehicle's infotainment system controls.

Cameras were mounted over each side-view mirror and pointed at the ground to provide visual indications of lane position and instances of lane departure. Lane departure data was not available for Visit 1; thus, Visit 1 is excluded from the analyses below.

Procedure

During each visit, participants completed 12 laps around the 2-mile test track following a lead vehicle driven by an experimenter. LDW was activated on half the laps. On some of these laps, participants also utilized the vehicle's Adaptive Cruise Control system. Data from Adaptive Cruise Control laps (6 of 12 for each visit) were not included in the analyses presented here. Participants were aware of the LDW condition of each lap. During each lap, participants followed a lead vehicle that controlled the pace and driving route. The lead vehicle varied its speed between 40 and 50 mph on the straightaways of all laps. Participants were asked to maintain a target speed of 40-50 mph at a comfortable following distance behind the lead vehicle, and to maintain a normal lateral position within the painted lane. Each visit was completed in approximately 50-60 minutes.

On select laps (counterbalanced across LDW conditions), participants performed a cognitive task concurrently with the driving task. During the straightaway sections of the laps, the experimenter read simple multiplication problems (e.g., 5×6, 6×8) to the participant. Participants were allowed to take as much time as desired to answer each question. Once an answer was provided, it was recorded and the experimenter asked the next question, continuing until the end of the straightaway. Each participant performed the concurrent cognitive task on only one of the three visits.

Participants were told that the purpose of the study was to explore how drivers responded to and felt about ADAS technologies. The specific performance measures used for analysis were not disclosed, nor was the purpose of the cognitive task.

During Visit 1, participants completed a questionnaire with questions related to demographic information and driving experience. Participants were also verbally introduced to the ADAS technologies in the vehicle prior to beginning the first drive. Participants then completed three practice laps to acclimate to the testing vehicle and ADAS. At the completion of all on-track driving on Visit 3, participants completed a questionnaire with questions related to their perceptions of ADAS technologies and experience with various driving situations. The results from this survey are discussed in detail in [23].

Data Processing & Analysis

LANE KEEPING

Lane departure events were coded by visual inspection of video footage of the front wheel positions within the lane (see example in Figure 2). Either the left or right front wheel crossing to the outer edge of a lane was considered a lane departure. The number of lane departures occurring within each lap was recorded. The duration of each lane departure was recorded as the time elapsed between the initial line crossing and the wheel re-entering the lane (i.e., no longer touching the lane line). Duration was recorded to the nearest second, with a minimum duration of one second. Total time of lane departures was calculated by summing the lane departure durations within each lap.

Due to a technical issue, lane departure data were unavailable for all participants at Visit 1, seven participants at Visit 2, and two participants at Visit 3. The following analyses were therefore conducted using data from those participants who had valid data at Visit 2 or Visit 3. At each visit, participants completed three laps in each of the two LDW conditions, yielding a total of 78 and 108 observations available for analysis from Visits 2 and 3, respectively Only data from the laps in which no secondary task was performed were included in these analyses in order to control for any potential effect of the secondary task on lane-keeping behavior.

Analyses were conducted on the counts and sum of durations of lane departures for all laps using Stata 14.1. Lane keeping data were first examined using histograms to identify outliers and assess normality. Measures of lane keeping (number of lane departures and total time of lane departures for each lap) were positively skewed.

FIGURE 2 Example frame of video used to determine lane departure events.

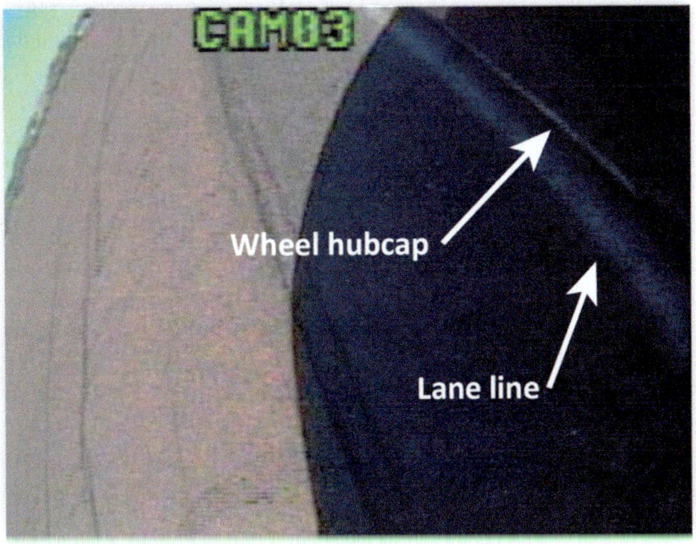

Stata program nbvargr was used to compare the data to Poisson and negative binomial distributions to determine the best fitting model. A Poisson model was used for number of lane departures and a negative binomial model was used for time of lane departures. Data were analyzed using multi-level models to account for nesting of observations within participants and because multi-level models effectively handle missing data by including all available observations in the analysis. LDW condition and Visit were included at level 1 and Participant at level 2. Models including a random slope and intercept were compared to models including only a random intercept using likelihood ratio tests. Inclusion of a random slope did not improve model fit, and therefore final models included only a random intercept. Models examining the lane keeping behavior dependent variables included the following predictors: LDW (0 = No LDW, 1 = LDW), Visit (2 = Visit 2, 3 = Visit 3), LDW × Visit interaction, and lap number (as a covariate).

COGNITIVE LOAD

Five participants completed the cognitive task at Visit 1, five completed the cognitive task at Visit 2, and ten completed the cognitive task at Visit 3 (cognitive task performance data was not available for one subject at Visit 3 due to an oversight, however). Each participant completed the cognitive task on one lap in each of the two LDW conditions, yielding a total of 10 and 18 observations available for analysis from Visits 2 and 3, respectively. Cognitive task performance was measured by the error rate on the multiplication task. Error rate was calculated as the number of incorrectly answered problems out of the total asked for each lap. To compare error rates, the Wilcoxon sign-rank test for matched-pairs was used due to non-normality of the data. Although data were available for Visit 1, error rates were compared between LDW laps and non-LDW laps for Visits 2 and 3 only, to allow some comparison with lane departure data for the same Visits.

Results

Lane Keeping

For number of lane departures, a significant LDW × Visit interaction (b = .71; SE = .31; 95% CI = .10 to 1.32; p = .023) was observed. Tests of simple effects revealed that at Visit 2, participants departed the lane less frequently when the LDW system was active than when it was not (difference = −.53; SE = .24; 95% CI = −1.01 to −.06; p =.029), whereas at Visit 3, the number of lane departures did not significantly differ by LDW condition (p = .359). No significant change in number of lane departures was observed from Visit 2 to Visit 3 in either LDW condition (p > .059) (Figure 3).

Similarly, for duration of lane departures, there was a marginally significant LDW × Visit interaction (b = .75; SE = .38; 95% CI = .00 to 1.49; p = .050). Tests of simple effects revealed that at Visit 2, participants spent less total time out of the lane with LDW than without (difference = −.72; *SE* = .29; 95% CI = −1.30 to −.14; p = .014), and at Visit 3, duration of lane departures did not significantly differ by LDW condition (p = .912). In addition, the duration of lane departures increased from Visit 2 to Visit 3 in the LDW condition (difference = .85; SE = .30; 95% CI = .27 to 1.44; p = .004), but not in the no-LDW condition (p = .702) (Figure 4).

FIGURE 3 Effect of LDW and Visit on number of lane departures. The *y* axis represents the mean number of lane departures per lap predicted by the model.

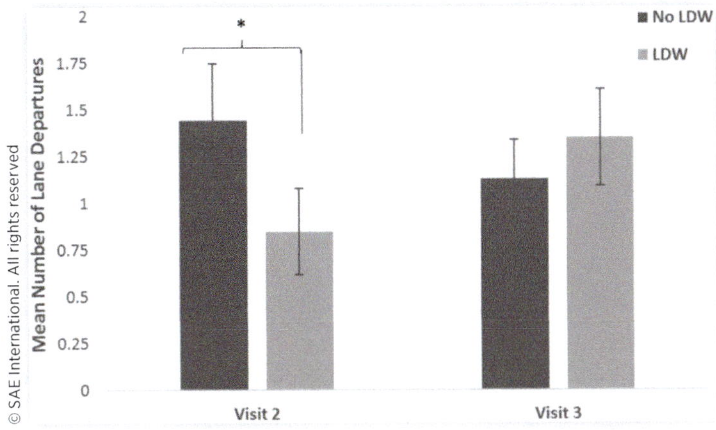

FIGURE 4 Effect of LDW and Visit on duration of lane departures. The *y* axis represents the mean duration of lane departures predicted by the model.

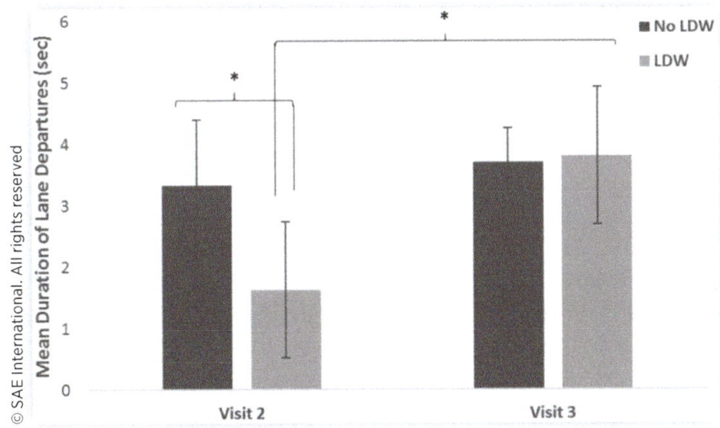

Cognitive Task Performance

Higher error rates on the cognitive task were treated as an indicator of greater cognitive load associated with the concurrent driving task. Because participants completed the cognitive task on only one of the three visits, we first examined the main effect of LDW on cognitive task performance on Visits 2 and 3 combined. Although the mean error rate was higher in the LDW condition, a significant difference was not observed ($z = 1.57$, $p = .116$).

Cognitive performance was also analyzed separately for Visits 2 and 3. The results showed that performance on the concurrent cognitive task did not significantly differ by LDW condition for Visit 2 ($z = .41$; $p = .686$). At Visit 3, however, performance was significantly worse (error rate was higher) when LDW was in use ($M = .11$; $SD = .08$) compared to when it was not ($M = .03$, $SD = .05$; $z = 2.31$; $p = .021$) (Figure 5).

FIGURE 5 Effect of LDW on cognitive task performance at Visit 3.

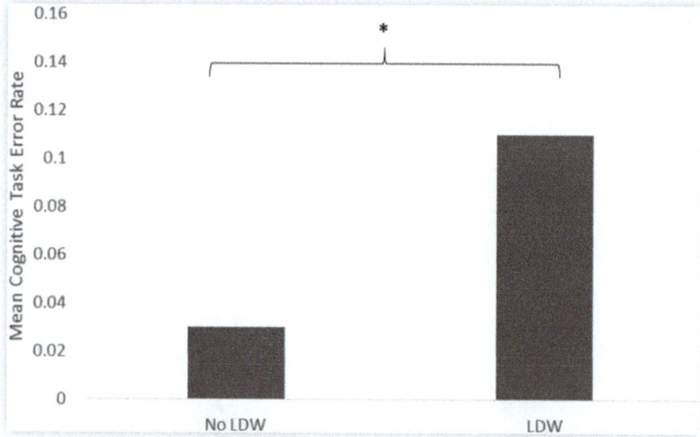

Discussion

The results showed that use of LDW improved lane-keeping (as measured by the number of lane departures and the total time of lane departures), but only in the earlier visit analyzed (Visit 2). In fact, during Visit 3, no effect of LDW was observed on either measure. This result partially supports the first part of our hypothesis, that LDW would improve drivers' ability to maintain an appropriate lateral position.

A potential explanation for the apparent reduction in effectiveness of LDW over time in our results is that drivers underwent behavioral adaptation as a result of their experience with LDW during the course of the study. Since the majority of our participants were unfamiliar with LDW before the study, they may have experienced the alerts as novel and salient in the earlier visits, but begun to "tune out" the alerts by Visit 3. This idea is supported by the fact that for both measures of lane keeping performance, the difference across visits appears partially or entirely driven by a worsening of performance in the LDW condition (see Figures 3 and 4). Such a finding appears to be consistent with at least one previously published study [10] in which severity of driving error was found to increase over time among participants driving with LDW.

The second part of our hypothesis, that LDW would impair performance on the cognitive task, was also partially supported by our data. We found that performance on the cognitive task was worse in the LDW condition, but only at Visit 3. However, it is difficult to evaluate the extent to which this result is associated with driving behavior (i.e., lane keeping performance), as this was after lane-keeping behavior appeared to habituate to the effect of LDW. Our ability to interpret the lack of such an effect in relation to the lane-keeping benefits at Visit 2 is additionally limited by a lack of overlap in the participants included in the lane-keeping and cognitive performance analyses for that visit (lane-keeping data was only available for one of the five participants who performed the cognitive task at Visit 2), yet there was more overlap between participants in the analyses for Visit 3 (lane-keeping data was available for seven of the nine participants for whom cognitive task performance data was also available at Visit 3). Thus, the significant effect of LDW on cognitive task performance that we observed at Visit 3 suggests a potential relationship between LDW, cognitive load, and driving performance, although the nature of this relationship remains unclear. Our data imply that it is not a simple shifting of cognitive resources to the lane-keeping task as hypothesized, since there was no corresponding improvement in lane-keeping with LDW for Visit 3. A prior analysis of other driving performance data, published in [20], suggested that the combination of LDW and performance of a cognitive task negatively impacted longitudinal vehicle control (i.e., it increased variability in speed) when compared to performance of the cognitive task without LDW. Thus, a shift in available cognitive resources may occur with the use of LDW, but may affect different elements of the driving task (such as longitudinal control) than those measured analyzed here.

Cognitive science research has frequently used performance on various concurrent cognitive tasks as a proxy for the amount of free attentional resources that remain while a participant is performing a primary task which also requires such resources. In this paradigm, reduced secondary task performance is generally interpreted to indicate reduced available attention and/or cognitive resources due to the demands of the primary task. Taken in this light, the reduction in secondary task performance reported here may indicate a related increase in cognitive load associated with the primary task (driving, or more specifically, lane-keeping). A potential mechanism for such an increase in cognitive requirements may be that LDW increases the "difficulty" of lane-keeping by enforcing a stricter lateral position requirement than would be considered acceptable by an attentive driver relying only on visual monitoring of the road to control lane position. A comparison of the LDW criteria for lane departure to the amount of deviation judged by drivers to require a lateral correction could be explored by a future study.

Our results conflict to some degree with existing literature on the effect of ADAS on cognitive load, which has generally reported improvements on secondary task performance with use of ADAS. However, as discussed in the Introduction, earlier studies in this realm mainly focused on ADAS technologies in which automated systems both detect and correct undesirable driving conditions (e.g., Adaptive Cruise Control, Active Steering), while LDW performs only the first function, requiring a response from the driver to accomplish the second. For this reason, LDW and other types of ADAS which provide feedback to the driver but do not intervene in the driving task (e.g., driver sleep detection systems) may represent a qualitatively distinct type of human-machine interaction for which the effects on driver behavior should be considered separately.

For a system such as LDW which relies on driver action rather than autonomous vehicle response to correct problematic driving behavior, driver attention to the driving task necessarily must be increased (at least at certain critical moments) in order to produce any actual change in driving performance compared to non-assisted driving. While such a diversion of drivers' cognitive resources toward lane-keeping (and away from any other potentially competing cognitive activities such as conversation, operation of entertainment or communication devices, or simply contemplation) may be theoretically desirable, it is important to consider whether drivers actually perceive the process as beneficial, neutral, or detrimental to their personal goals and comfort while driving. Research has shown that acceptability and effectiveness of driver warning systems are not necessarily correlated, and are sometimes contradictory [22]. If, from a driver's perspective, LDW interferes with the ability to perform the types of cognitive tasks mentioned above (which the driver may wish to perform regardless of any potential negative effect on driving performance) and does not provide an observable benefit that outweighs the perceived negative interference from the driver's perspective, acceptance will be low. Drivers may attempt to circumvent LDW by disabling it or opting for vehicles without it.

Indeed, several studies indicate that driver acceptance of LDW is lower than other ADAS technologies [11, 12, 13]. Analysis of the post-experiment survey data in this study indicates that 30% of subjects reported they would turn LDW off-more than those who would deactivate forward collision warning, auto-braking, or adaptive cruise control (see [23] for further discussion and analysis of the questionnaire results). Our finding that the lane-keeping benefits of LDW decline over time is also consistent with the possibility that drivers will not always respond to the technology in a way that actually improves driving safety. It could be that drivers do not perceive slight lane departures to be as dangerous as other driving events such as forward collisions and are thus less accepting of technology in this arena that interrupts their driving and performance of secondary tasks. For example, previous studies we conducted found both a performance benefit and general acceptance of passive collision warnings in the context of a frontal collision scenario [21, 23]. In that study, participants (in the same vehicle, on the same test track, following a balloon vehicle) were required to respond to a simulated sudden lead-vehicle deceleration which interrupted performance of a secondary cognitive task (mental math similar to the task reported here) in half the participants. Those participants completing the cognitive task benefitted from the presentation of the forward collision warning in that they were able to respond more quickly to the emergency event, relative to those who did not experience the forward collision warning. Post-experiment surveys further indicated that those participants who performed the cognitive task and experienced the forward collision warning rated the forward collision system highly [23]. Taken together, the results from [21] and the present results suggest that more passive warning systems may be acceptable "interrupters" only in the context of perceived emergencies, and not during less severe scenarios such as a lane departure with little or no surrounding traffic.

The current study is preliminary in nature and has some limitations that should be noted. First, due to the data presented here having been collected as one component of a larger, more generalized research effort, our participant sample size utilized for the present analysis was relatively small, and the cognitive task was performed on just one of the three visits for each participant. Further, unequal cell sizes in our dataset as a result of missing data limit the power to detect significant effects and our overall capability to generalize findings to a larger sample. Second, although having participants drive on a test track may be a better approximation of real world driving behavior than a driving simulator, the track still creates an artificial driving environment that may sometimes produce behavior that differs from that in real-world driving settings. Third, although we observed an apparent decrease in the effectiveness of LDW on lane-keeping behavior from Visit 2 to Visit 3, an examination of performance at more than two points in time would allow a more detailed characterization of any possible behavioral adaptation effects. Finally, the math problems used were relatively easy. Utilizing more difficult problems in the future may require greater cognitive resources, leading to more pronounced effects. We hope that the findings reported here will be further tested and expanded upon by both our own future research as well as research by others, which may yield more definitive conclusions.

Nevertheless, the current findings show that lane departure warnings are able to effectively improve lane keeping, but may wane in effectiveness over time. In addition, our findings suggest that LDW alerts are associated with impairments in performance on a secondary cognitive task, possibly because they increase the cognitive load associated with maintaining lateral vehicle control. This may partially explain why drivers have demonstrated less acceptance for LDW than for other types of ADAS. These observations lend credence to the notion that a more complete transition to fully autonomous driving systems, rather than to partially or highly autonomous systems, may ultimately be more acceptable to drivers. Although fully autonomous systems will have their own challenges to overcome in terms of public perception and trust, the potential safety benefits of ADAS technologies may be more fully and expediently realized if potential conflicts between automated systems and driver goals and preferences are avoided.

Contact Information

Helene G. Moorman, Ph.D.
Exponent, Inc.
149 Commonwealth Dr.
Menlo Park, CA 94025
hmoorman@exponent.com

References

1. Bengler, K., Dietmayer, K., Farber, B., Maurer, M. et al., "Three Decades of Driver Assistance Systems: Review and Future Perspectives," *IEEE Intell. Transp. Syst. Mag.* 6, no. 4 (2014): 6-22.
2. "Predicted Availability of Safety Features on Registered Vehicles—An Update," *Highw. Loss Data Inst. Bull.* 31, no. 5 (2014).
3. Paul, A., Chauhan, R., Srivastava, R., and Baruah, M., "Advanced Driver Assistance Systems," SAE Technical Paper 2016-28- 0223, 2006, doi:10.4271/20016-28-0223.
4. NHTSA, "Traffic Safety Facts 2014: A Compilation of Motor Vehicle Crash Data from the Fatality Analysis Reporting System and the General Estimates System," DOT HS 812 261, 2014.
5. Sternlund, S., Strandroth, J., Rizzi, M., Lie, A. et al., "The Effectiveness of Lane Departure Warning Systems—A Reduction in Real-World Passenger Car Injury Crashes," *Traffic Inj. Prev.* 0 (2016), doi:10.1080/15389588.2016.1230672.
6. LeBlanc, D., Sayer, J., Winkler, C., Ervin, R., et al., *Road Departure Crash Warning System Field Operational Test: Methodology and Results. Volume 2: Appendices* (2006).
7. Deroo, M., Hoc, J.-M., and Mars, F., "Influence of Risk Expectation on Haptically Cued Corrective Manoeuvres during Near Lane Departure," *Ergonomics* 55, no. 4 (2012): 465-475.
8. Navarro, J., Yousfi, E., Deniel, J., Jallais, C. et al., "The Impact of False Warnings on Partial and Full Lane Departure Warnings Effectiveness and Acceptance in Car Driving," *Ergonomics* (2016): 1-12, doi:10.1080/00140139.2016.1158323.
9. Rimini-Doering, M., Altmueller, T., Ladstaetter, U., and Rossmeier, M., "Effects of Lane Departure Warning on Drowsy Drivers' Performance and State in a Simulator," October 2016.
10. Kozak, K., Pohl, J., Birk, W., Greenberg, J. et al., "Evaluation of Lane Departure Warnings for Drowsy Drivers," *Proceedings of the Human Factors and Ergonomics Society Annual Meeting*, Sage Publications, 2006, 2400-2404.
11. "Luxury Owners Embrace Systems to Skirt Crashes," *IIHS Status Rep.* 44, no. 10 (2009).
12. Sanchez, D., Garcia, E., Saez, M., Benmimoun, M. et al., "European Large-Scale Field Operational Tests on In-Vehicle Systems," *EuroFot* (2012).
13. Reagan, I.J. and McCartt, A.T., "Observed Activation Status of Lane Departure Warning and Forward Collision Warning of Honda Vehicles at Dealership Service Centers," *Traffic Inj. Prev.* 17, no. 8 (2016): 827-832, doi:10.1080/15389588.2016.11 49698.

14. Ranney, T.A., "Models of Driving Behavior: A Review of Their Evolution," *Accid. Anal. Prev.* 26, no. 6 (1994): 733-750.

15. Dinges, D., Maislin, G., Brewster, R., Krueger, G. et al., "Pilot Test of Fatigue Management Technologies," *Transp. Res. Rec. J. Transp. Res. Board* (1922): 175-182, 2005.

16. Young, M.S. and Stanton, N.A., "Malleable Attentional Resources Theory: A New Explanation for the Effects of Mental Underload on Performance," *Hum. Factors* 44, no. 3 (2002): 365-375.

17. Merat, N., Jamson, A.H., Lai, F.C.H., and Carsten, O., "Highly Automated Driving, Secondary Task Performance, and Driver State," *Hum. Factors* 54, no. 5 (2012): 762-771.

18. Ma, R. and Kaber, D.B., "Situation Awareness and Workload in Driving While Using Adaptive Cruise Control and a Cell Phone," *Int. J. Ind. Ergon.* 35, no. 10 (2005): 939-953, doi:10.1016/j.ergon.2005.04.002.

19. Rudin-Brown, C.M. and Parker, H.A., "Behavioural Adaptation to Adaptive Cruise Control (ACC): Implications for Preventive Strategies," *Transp. Res. Part F Traffic Psychol. Behav.* 7, no. 2 (2004): 59-76, doi:10.1016/j.trf.2004.02.001.

20. Cades, D.M., Crump, C., Lester, B.D., and Young, D., Driver Distraction and Advanced Vehicle Assistive Systems (ADAS): Investigating Effects on Driver Behavior, Stanton, N.A., Landry, S., Bucchianico, G.D., and Vallicelli, A. eds., *Advances in Human Aspects of Transportation* (Springer International Publishing, 2017), 1015-1022, ISBN:978-3-319-41681-6, doi:10.1007/978-3-319-41682-3_83.

21. Crump, C., Cades, D., Rauschenberger, R., Hildebrand, H. et al., "Driver Reactions in a Vehicle with Collision Warning and Mitigation Technology," SAE Technical Paper 2015-01-1411, 2015, doi:10.4271/2015-01-1411.

22. El Jaafari, M., Forzy, J.F., Navarro, J., Mars, F. et al., "User Acceptance and Effectiveness of Warning and Motor Priming Assistance Devices in Car Driving," *Proceedings of European Conference on Human Centred Design for Intelligent Transport Systems*, 2008, 311.

23. Crump, C., Cades, D., Lester, B., Reed, S. et al., "Differing Perceptions of Advanced Driver Assistance Systems (ADAS)," *Proceedings of the Human Factors and Ergonomics Society Annual Meeting* 60, no. 1 (2016), SAGE Publications.

CHAPTER 2

Study on Test Scenarios of Environment Perception System under Rear-End Collision Risk

Lin Liu, Xichan Zhu, and Zhixiong Ma
Tongji University

The foundation of both advanced driving assistance system (ADAS) and automated driving (AD) is an accurate environment perception system (EPS). However, evaluation and test method of EPS are seldom studied. In this paper, naturalistic driving environment was studied and test scenarios for EPS under rear-end collision risk were proposed accordingly. **To describe driving environment, a new concept named environment perception element (EPE) was first proposed in this paper, which refers to all the objects that the EPS must perceive during driving. Typical environment perception elements include weather and light conditions, road features, road markings, traffic signs, traffic lights, other vehicles, pedal cyclists and pedestrians and others**. Driving behaviors collected in Shanghai, China were classified and rear-end collision risk scenarios were obtained and described using EPEs. **Probability distribution of EPEs was therefore obtained. Afterwards, the correlation between EPEs and risk level of scenarios (evaluated by maximum longitudinal deceleration) were revealed by means of Fisher's Exact Tes**t. Based on these two characteristics of driving environment, typical test scenarios for EPS were established with the help of cluster analysis, and the test scenarios were simulated in Prescan. These test scenarios were generally consistent with the probability distribution of EPEs, making the test results reliable. **Results from this paper fill the gap between the high demand of dynamic and representative EPS test scenarios and the existing static picture database used in development and validation of EPS and are of great significance to the development of ADAS and AD in China**.

CITATION: Liu, L., Zhu, X., and Ma, Z., "Study on Test Scenarios of Environment Perception System under Rear-End Collision Risk," SAE Technical Paper 2018-01-1079, 2018, doi:10.4271/2018-01-1079.

Introduction

Advanced driving assistance system (ADAS) and automated driving (AD) are playing a more and more prominent role in the development of vehicle technology. Some ADAS have already been put into mass production, such as Pilot Assist and City Safety in Volvo S90, Side Blind Zone Alert (SBZA) and line change assistant (LCA) in Buick Envision. Several AD systems have already finished some road test, including Drive Me of Volvo and Google's Driverless Car.

The basic premise of both ADAS and AD is an accurate perception and understanding of the surrounding environment of the vehicle, which can only be achieved by an effective environment perception system (EPS) [1]. So before testing ADAS or AD to confirm their safety benefits, EPS should be first evaluated.

In aspect of evaluation of widely used ADAS, only a small amount of foundational researches have been accomplished. The National Highway Traffic Safety Administration (NHTSA) has introduced regulations for AEB systems. United Nations Economic Commission for Europe (ECE) also released a draft regulation. In Europe, a few third-party testing organizations (such as ADAC, ASSESS etc.) outlined the test and evaluation scheme for AEB systems. Euro NCAP, for example, has added rating system for AEB, AEB City and AEB Interurban to the 2014 edition.

Figure 1 describes the CCRM (Car-to-Car Rear Moving) test scenario for AEB city. Target vehicle runs forward at 20 km/h and the test vehicle drives forward at 30-80 km/h behind target vehicle. During the test, working condition for AEB is simplified as the motion conflict between the test vehicle and the target vehicle. However, in the real driving condition, driving environment is much more complicated due to multiple vehicle targets and disturbance of road environment and pedestrians, which poses a great challenge for EPS. The performance of an EPS is uncertain in such a complex environment. Therefore, test scenarios, those can reflect the real and complex environment, have their rationality and necessity.

At present, the common test method of AD is field operational test (FOT) [2]. But according to Nidhi Kalra [3], hundreds of millions of miles or sometimes even hundreds of billions of miles have to be done to demonstrate the reliability of AD in terms of fatalities or injuries, which leads to high expenses and long test cycle. Furthermore, an untested vehicle means a potential threat to other human driving vehicles on roads. However, these two shortcomings can be overcome by virtual tests, in which evaluation process can be accelerated using representative test scenarios and no accidents would really happen. An important step towards virtual tests is the establishment of test scenarios database. To ensure reliability of tests on EPS of ADAS and AD, unbiased tests scenarios are of great importance, the establishment of which is the main objective of this paper.

Real driving process can be divided into a large number of different scenarios. Due to time and article length limitation, the rear-end risk scenarios were mainly considered

FIGURE 1 Test scenario CCRM for AEB city.

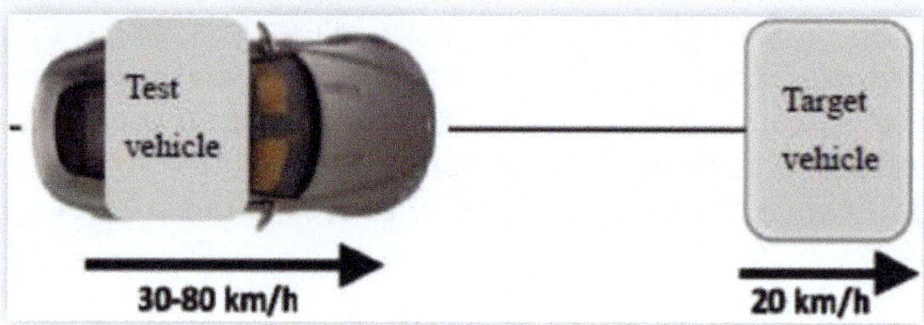

in this paper. According to a report from NHTSA, in 2010, 32% of all the 1.7 million accidents in American were rear-end collisions [4]. In China, rear-end collisions accounted for 34.29% in 2013 and led to serious casualties and loss of properties [5].

In this paper, test scenarios of EPS under rear-end collision risk conditions were proposed. Two naturalistic driving tests were conducted in Shanghai to guarantee the reliability of the proposed scenarios. The real driving data were collected and classified and then rear-end risk scenarios are obtained. **To describe the driving environment, environment perception element(EPE) was defined, which refers to all the objects that the EPS must perceive during driving, including weather and light conditions, road features, road markings, traffic signs, traffic lights, other vehicles, pedal cyclists and pedestrians and others.** Probability distribution of EPEs in rear-end risk scenarios was then obtained. At the meantime, correlation between EPEs and risk level of scenarios (evaluated by maximum deceleration) was exposed using Fisher's Exact Test. At last, typical test scenarios for EPS were acquired by cluster analysis and simulated in Prescan.

Materials and Processing

Extraction of Real Naturalistic Traffic Rear-End Risk Scenarios

In order to collect real traffic scenarios, two naturalistic driving tests were conducted in Shanghai. In the earlier four-year test, driving data of a total of 10 taxis and 3 police cars were recorded using video drive recorders (VDRs), which could read longitudinal velocity, acceleration of X, Y, Z direction of the test vehicle from CAN-bus. When the longitudinal acceleration of the test vehicle was greater than the given threshold, 0.4 g, driving data would be recorded, including traffic road images in front of the vehicle, driver behavior, speed, acceleration of X, Y, Z direction during the 20s (15 s before, 5 s after) interval before and after the trigger time, as shown in Figure 2.

The total travel length of all 13 vehicles was about 1.28 million kilometers and about 3000 cases were collected. All recorded videos were reviewed and the false positive ones were excluded. Finally, 500 risk scenarios, named as "VDR-500", were obtained.

FIGURE 2 Real risk scenarios in China.

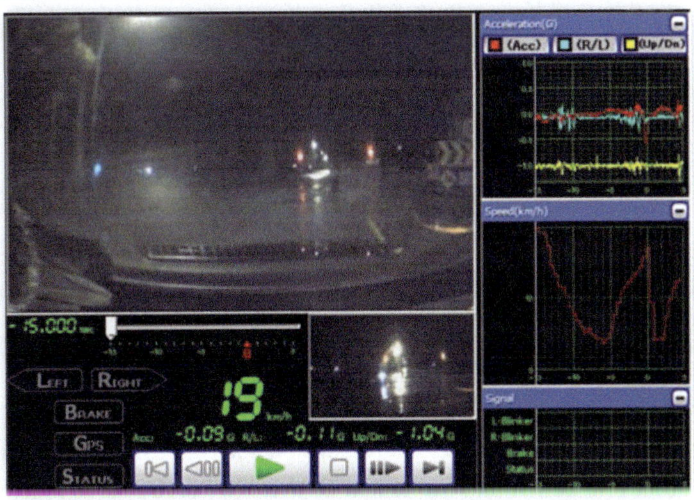

FIGURE 3 Absolute value of longitudinal deceleration.

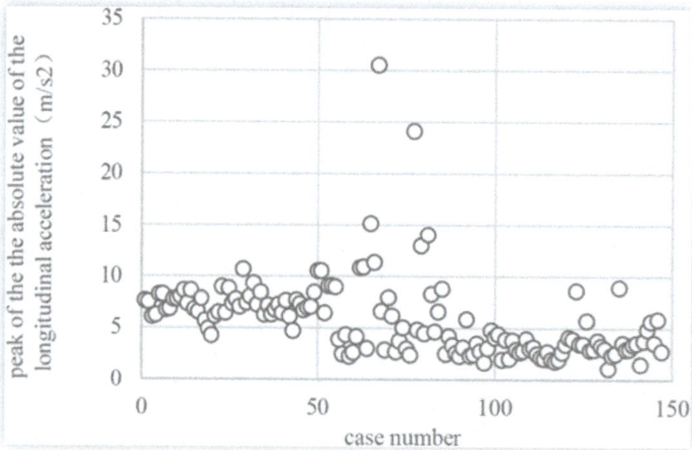

Since July 2014, an 18-mounth field operational test was conducted and more detailed and complete driving data were obtained. Road environment both in front and rear of vehicle, driver's behavior and pedal operation were recorded in form of video. At the meantime, data in CAN-bus of the vehicle were read and saved. By comparing the acceleration with the threshold, 179 risk scenarios, named "FOT-179" were finally extracted from a total of 7402 video clips, 3594 hours long video database.

In order to obtain rear-end risk scenarios, VDR-500 and FOT-179 were classified according to the pre-crash scenario typology defined by NHTSA [6]. There were 148 rear-end risk scenarios, which account for 21.80% of all the 679 near-crash cases. So rear-end risk scenarios are worth attention and related EPE research is essential. These 148 cases can be further divided into three types: the test vehicle approaches stationary, slower or braking target vehicle.

Study of Dangerous Level of Risk Scenarios

There are several methods to define the dangerous level of the risk scenarios, such as time to collision (TTC) and risk perception proposed by Kondoh [7]. In this paper, risk scenarios were obtained by comparing test vehicle's longitudinal acceleration with thresholds. Dangerous level is therefore classified using the acceleration. The longitudinal acceleration distributions of all 148 scenarios are shown in Figure 3. The vertical axis stands for the absolute value of the longitudinal acceleration. Horizontal axis represents the case number. Number 1-55 indicates FOT-179 and Number 56-148 indicates VDR-500.

The figure above shows that the acceleration is mainly distributed in the range of 2-10 m / s^2. Dangerous levels are evaluated using Table 1.

Methods

Environment Perception Elements (EPEs) Considered in Analysis

The environment perception system is responsible for collecting and comprehending the environmental information and predicting the changing trend of the environment [1].

TABLE 1 Dangerous level of risk scenarios

Absolute value of longitudinal acceleration(m/s2)	Dangerous level	Number of risk scenarios
$0.0 \le a < 2.5$	1	17
$2.5 \le a < 5.0$	2	57
$5.0 \le a < 7.5$	3	35
$7.5 \le a < 10.0$	4	27
$a \ge 10.0$	5	11

Driving environment is a general concept, which refers to the surroundings of a vehicle. Any changes of the properties of the surroundings will lead to interaction effects with the vehicle. In an ADAS or AD system, EPS has to understand all the environment variables that may have effects on the test vehicle. The concept, environment perception elements (EPE), is hereby put forward.

EPEs refer to the complete set of all environmental variables in a certain period of time and space, which are required to be observed and understood by EPS during driving. In real driving scenarios, the time range should cover the entire process of driving, even a certain time range before the vehicle starts. In principle, the more environment information the automated car gets, the safer decisions can be made by the decision system, hence, the scope of the space should be as large as possible. However, limited by the data source and workload, the time range in this paper refers to the time range of the data, that is, 10s before and 5 s after the longitudinal acceleration achieves the peak value. The spatial range refers to the road image available in the forward camera.

EPEs are first classified and summarized into eight categories, weather and light conditions, road features, road markings, traffic signs, traffic lights, other vehicles, pedal cyclist and pedestrians and others. Each category is subdivided into multiple Subprojects. For example, weather conditions include no precipitation, rainfall, snowfall, fog and gale. However, some kind of weather did not appear due to the limitation of natural conditions in Shanghai. Similar lack of subprojects exists also in other categories.

Probability Distribution of EPEs

Statistical work of EPEs was carried on in all 148 rear-end risk scenarios from two databases. In scenario No. 45_ GS081111_200058 and No. 41_GS090316_223832, the lighting condition is too weak, so the image could not be identified. Statistical results of the remaining 146 valid risk scenarios are as follows. Due to article length limitations, only the objects with the highest frequency under each EPEs are listed in the following tables.

WEATHER AND LIGHT CONDITIONS

Weather and light conditions may have a great effect on EPS' perception of environment. For example, the rain has different influence on camera and Radar. So, the weather condition should be perceive and the confidence level of sensors in sensor-fusion should be changed accordingly.

Probability distribution of weather and light conditions is shown in Table 2. Due to the limitation of weather conditions in Shanghai, there were only two kinds of weather conditions, no precipitation and precipitation, with 137 cases and 9 cases and accounting for 93.88% and 6.12%, respectively.

TABLE 2 Probability distribution of weather and light conditions

EPEs	Description	Ratio (%)
Weather	No precipitation	93.88
	Rainfall	6.12
Lighting condition	Natural daylight	57.14
	Artificial light at night	30.61

57% of the risk scenarios are with natural daylight, and there is artificial light in the rest 30.61% scenarios at night. In particular, there is a scenario without light at night. It can be seen from the video image that it is difficult for the driver to recognize the environment object in this scenario. Such scenarios are relatively rare, but they pose a great challenge to the environment perception system.

ROAD FEATURES AND ROAD MARKINGS

In real driving process road environment is very complex. In order to describe it in detail, the structure of the road is decomposed into the following EPEs. Classification of road marks mainly refers to China's national standard, 'GB 5768.3-2009 Road traffic signs and markings Part 3 Road traffic markings' [10]. Probability distribution is described in Tables 3 and Table 4.

TRAFFIC SIGNS

Classification of traffic lights mainly refers to China's national standard, 'GB 5768.3-2009 Road traffic signs and markings Part 2 Road traffic signs' [10]. Probability distribution is described in Table 5.

TABLE 3 Probability distribution of road features

EPEs	Description	Ratio (%)
Left shoulder	Bush	43.75
	Hard shoulder	34.82
Right shoulder	Bush	41.3
	Hard shoulder	25.17
Surface water	Dry	93.15
	Surface water	6.85
Surface roughness	flat	90.41
	bumps	9.59
Shape of the road	straight	85.81
	curve	14.19
Type of roads	Normal road	77.84
	On the bridge	8.38
Gradient	horizontal	95.95
	Uphill	2.03
Number of lanes	Two-way four-lane	55.24
	Two-way two-lane	12.59
Lane separation	Road marks	51.11
	Fence	25.93
Special Road	crossroad	65.77
Structures	T-Junction	24.32

TABLE 4 Probability distribution of road markings

EPEs	Description	Ratio (%)
Informative marking	dashed white lines	32.01
	zebra crossing	23.51
Prohibitory marking	Stop line	46.58
	Solid middle lines	34.93

TABLE 5 Probability distribution of traffic signs

EPEs	Description	Ratio (%)
warning signs	zebra crossing ahead	62.47
	crossroads ahead	11.82
Prohibitory signs	Speed limits	60.53
	No right, left, or U-Turn	10.53
Mandatory sign	Proceed straight	30.43
	Segregated lane	30.43
Direction sign	In urban road	83.33
	In Highway/City Expressway	16.67

TRAFFIC LIGHTS

Signal lights are important information that the vehicle needs when driving through intersections. The classification of traffic lights mainly refers to "GB14886-2006 Setting and installation specifications of Road Traffic Signal " [11] and "GB14887-2011 Road Traffic Signal Lights" [12]. Probability distribution is described in Table 6.

OTHER VEHICLES, PEDAL CYCLISTS AND PEDESTRIANS

Probability distribution is described in Table 7.

OTHERS

The EPEs that can not be attributed to the first six categories are summarized in this part. Gestures of police, for example, especially when there is a difference between police'

TABLE 6 Probability distribution of weather and light conditions

EPEs	Description	Ratio (%)
Traffic lights	Traffic lights for vehicles	72.73
	Traffic lights for pedestrians	22.22

TABLE 7 Probability distribution of other vehicles, Pedal cyclists and pedestrians

EPEs	Description	Ratio (%)
Vehicles in the same direction	car	75.77
	van	10.86
Vehicles in the opposite direction	car	79
	van	10.23
Types of pedal cyclists	Electric bike	85.63
	bicycle	11.14
Types of pedestrian	minor	1.09
	adult	98.91

gesture and traffic lights, are very important for drivers. Some obstacles may appear during driving, such as animals, large stones, etc., which the vehicle must identify. Special constructions on roads are also classified, parking fees and toll stations and so on. However, in all the 148 risk scenarios, these EPEs are rare.

Study of the Correlation between EPEs and Dangerous Level of Risk Scenarios

The correlation between the EPEs and dangerous level shows the importance of different EPEs. For example, when the analysis results show that the weather conditions are closely related to the dangerous level and a high-risk condition always occurs under rainy weather conditions, then it is necessary to set a rainy weather scenario in evaluation process.

Chi-square test is a non-parametric test, mainly used for the correlation analysis of two categorical variables. The basic idea is to test whether the actual frequency coincident with the theoretical frequency. However, in practice, the frequency of some EPEs is too low, leading to false test results. Fisher's exact test, which can be employed on small size samples, is therefore used. **Fisher's exact test, named after its inventor Ronald Fisher, is a statistical significance test used in the analysis of contingency tables [13]. Although in practice it is employed when sample sizes are small, it is valid for all sample sizes.**

Fisher's method was used to analyze the correlation between EPEs and dangerous level of risk scenarios. **The results are shown in the following Table 8. T-value is the calculate result of t- test and the Sig. is the significance of t-value. At a significance level of 0.05, 11 EPEs in Table 8 show great correlation with dangerous level (Sig. <0.05)**, including the weather conditions, right shoulders, types of roads, lane separation, vehicles in the opposite direction, pedestrian's location, posture and movement, pedal cyclists' location, posture and movement.

Generation of Test Scenarios for EPS

The features of EPEs obtained in this paper mainly include two aspects, the frequency characteristics and the correlation with dangerous level. Based on these two

TABLE 8 Result of Fisher's exact test

			Monte Carlo Sig.		
				99% confidence interval	
Classification of EPEs	EPEs (variable)	t-value	Sig.	Lower limits	Upper limits
weather and lighting conditions	Weather	8.958	.022	.019	.026
Road and road Marks	Right shoulder	29.943	.025	.021	0.029
	Type of roads	28.981	.002	.001	.004
	Lane separation	34.745	.025	.021	.028
the other vehicles	Vehicles in the opposite direction	10.201	.039	.034	.044
Pedal cyclists and pedestrians	Pedestrians' location	31.253	.000	.000	.000
	Pedestrians' posture	10.615	.027	.022	.031
	Pedestrians' movement	66.626	.000	.000	.000
	Pedal cyclists	31.970	.001	.000	.002
	Pedal cyclists' location	26.677	.020	.016	.024
	Pedal cyclists' movement	80.636	.000	.000	.000

characteristics, test scenarios for EPS will be obtained by means of cluster analysis. Cluster analysis is a method to define subgroups of a population, which has been widely used in transportation area [8, 9].

According to the statistical results, the EPEs are divided into two groups. In the first group, the element appears in each scenario and appears only once, for example, there is a weather condition in each scenario. In the cluster analysis, these elements are defined as interval scale variables. Another type of EPEs, nominal scale variables, can be described in terms of quantities, such as the number of pedestrians or the number of other vehicles.

The cluster analysis used in this paper is system cluster analysis. At the beginning of analysis, each of the 148 risk scenarios are taken as a separate cluster. First of all, the distance between the different sample variables is calculated. For the interval scale variables, the distance is defined as the absolute value of the difference (normalized to 0-1). For a nominal scale variable, the distance is 0 between two identical variable values and 1 between two different ones. After the distance between different sample variables is obtained, the distance between the different samples can be calculated. In this paper, we use the absolute distance to quantify the distance between samples. Set 148 samples of risk scenarios as X_1, $X_2 \ldots X_{148}$, then the distance between the scenario i-th and the scenario j-th is defined as the sum of the distances between two sample variables:

$$d_{ij} = \sum_{k=1}^{n} \left| x_{ik} - x_{jk} \right| \tag{1}$$

According to the distance between samples, distance between clusters can be calculated, where the average method is used. The average distance between clusters is defined as the average of the squared distances between samples. Assume that cluster G_K have n_K samples and $G_L$class have n_L samples, then the square distance between cluster G_K and G_L is:

$$D_{KL}^2 = \frac{1}{n_K n_L} \sum\nolimits_{x_i \in G_K, x_j \in G_L} d_{ij} \tag{2}$$

The closest two clusters, between which the distance is smallest, are merged into a new cluster. For the acquired 147 clusters, the above-mentioned actions are repeated until the best classification is obtained. The results of cluster analysis will be described in detail in the results.

Results

In the software SPSS [14], the risk scenarios were clustered into 4-10 categories. In order to determine the most appropriate number of clusters, comparisons between variables were tested by One - Way ANOVA analysis in each cluster method. And results showed that, significant difference between variables was found when the dataset was divided into 9 clusters. The chi-square test was carried out on nominal scale variables (on 90% confidence level), that is, comparing the proportion of each variable in each cluster with the proportion of the variable in all scenarios. Typical characteristics of each kind of risk scenario were then obtained. Finally, three typical test scenarios were obtained, shown in Table 9. They represent for 91.67% of all the risk scenarios.

The test scenarios obtained in the previous section were then modeled in Prescan with idealized data to visualize the scenarios, as shown in Figures 4 and 5. It's the basis for virtual test of EPS. Prescan is a physics-based simulation platform that is used in the automotive industry for development of ADAS from TASS International [15]

TABLE 9 Test Scenarios for EPS

EPEs	Cluster 1	Cluster 2	Cluster 3
Number	52	70	10
Ratio (%)	36.11	48.61	6.94
weather	No precipitation	No precipitation	No precipitation
lighting condition	Natural daylight	Natural daylight	Natural daylight
left shoulder	Bush	Bush	Hard shoulder
right shoulder	Bush	Bush	Bush
surface water	Dry	Dry	Dry
type of road	Normal road	Normal road	On the bridge
Gradient	horizontal	horizontal	horizontal
surface roughness	flat	flat	bumps
special road structures	crossroad	null	crossroad
number of lanes	Two-way four-lane	Two-way four-lane	Two-way four-lane
lane separation	Road marks	Road marks	Bush
shape of roads	straight	straight	straight
road marks	Standard traffic marks	Standard traffic marks	Standard traffic marks
informative marking	3	2	4
prohibitory marking	1	1	1
warning signs	0	0	0
prohibitory sign	0	0	1
mandatory sign	0	0	0
direction sign	0	0	1
traffic light for vehicles	1	0	1
traffic lights for pedestrians	0	0	1
vehicles in the same direction	8	4	2
vehicles in the opposite direction	5	5	4
pedestrian	1	1	1
pedal cyclist	3	2	1

FIGURE 4 Test scenarios for EPS (Cluster 1, Pilot view).

FIGURE 5 Test scenarios for EPS (Cluster 1, south direction).

Figure 4 and 5 depict the modelling results of cluster 1 mentioned above. Figure 4 is pilot view and Figure 5 is from south direction. The virtual scenarios can be used for evaluation of EPS by means of hardware in loop method with test vehicles.

Discussion

The main object of this paper is the establishment of test scenarios for environment perception systems. Before solving this problem, real driving environment data must be obtained and the characteristic of the environment must be learned.

First, two naturalistic driving tests were conducted in Shanghai and real risk scenarios were obtained. In Ding Zhao's paper, data collected in the Safety Pilot Model Deployment (SPMD) project in American was used for accelerated evaluation of AD [16]. Due to the different traffic environment in various countries, this evaluation method may not be suitable for China. So the environment information in this paper is supportive for the localization of ADAS.

Then characteristic of the environment was described and analyzed using concept EPEs. A detailed probability model of driving environment was then established, making virtual test on EPS possible. Afterwards, correlation between EPEs and dangerous level of risk scenarios was contained using Fisher's Exact Test.

At last, a set of test scenarios were proposed, which can be attached to the present test method of ADAS and AD. For example, In Mikael Ljung's framework for ADAS evaluation, collisions between vehicles and driver behavior were discussed in detail, while the driving environment is omitted. [17] Euro NCAP also simplifies test scenarios to conflicts between two vehicles, ignoring the EPEs' potential impact on ADAS. In addition, previous tests on EPS were limited to perception of one or several environment elements, such as the tests conducted by German Traffic Sign Recognition Benchmark (GTSRB) and INRIAPerson, in which only a database for pedestrian recognition was used [18]. Test scenarios proposed in this paper is in detailed and may pose more severe challenges on EPS to ensure the safety of ADAS and AD.

Some limitations of the present study should be mentioned. Firstly, due to limited data sources, the number of rear-end risk scenarios obtained in this paper is not large enough due to limited data sources. And the frequency of many EPEs is very low, so some EPEs did not show in the statistical result. In addition, naturalistic driving tests were mainly conducted in Jiading District, Shanghai. So the results are lack of special weather such as snow and other special road types such as mountain road. The results obtained are only representative for this region.

More work should be done to collect more scenarios in order to make the established test scenarios more reliable. And test scenarios under other types of risk will be contained. Since larger datasets will make the data more complex, an effort-saving analysis method of naturalistic driving data will be studied in the future.

Conclusions

In this paper test scenarios for environment perception system under rear-end collision risk were proposed. First of all, datasets 'VDR-500' and 'FOT-178' were obtained by means of naturalistic driving tests and then classified. A total of 148 rear-end risk scenarios were extracted, including three types, following a stationary, slower or braking front vehicle. At the same time, EPEs were defined as complete set of all environmental variables in a certain period of time and a certain scope of space. Its seven subcategories include weather and lighting conditions, road features, road markings, traffic signs, traffic lights, other vehicles, pedal cyclists and pedestrians and others. Through statistics analysis of the driving environment of all the 148 risk scenarios, the probability distribution of EPEs were obtained. After that, the correlation between EPEs and risk levels was analyzed using Fisher's Exact Test. The factors closely related to the risk levels are: weather conditions, right shoulders, types of roads, lane separation, vehicles in the opposite direction, pedestrians' location, posture and movement, pedal cyclists' location, posture and movement. Finally, a set of scenarios for testing is obtained using the Fisher's exact test.

This result can be used for hardware-in-the-loop test and virtual test of EPS of ADAS and AD systems, supporting the development processes of EPS.

Contact Information

Lin Liu
imshirleyll@163.com
School of Automobile Studies, Tongji University,
Shanghai, China

Xichan Zhu
zhuxichan@tongji.edu.cn
School of Automobile Studies, Tongji University,
Shanghai, China

Acknowledgments

This work is supported by the National Key R&D Plan of China (No. 2016YFB0100904-2).

References

1. Endsley, M.R. and Garland, D.J., "Theoretical Underpinnings of Situation Awareness: A Critical Review," *Situation Awareness Analysis and Measurement* (2000): 3-32, https://www.researchgate.net/publication/230745477_Theoretical_underpinnings_of_situation_awareness_A_critical_review.
2. Consortium, F., "FESTA Support Action Field opErational teSt supporT Action. D2.1—A Comprehensive Framework of Performance Indicators and Their Interaction," © festa 2008, 2008, https://dspace.lboro.ac.uk/dspace-jspui/handle/2134/5701.
3. Kalra, N. and Paddock, S.M., "Driving to Safety: How Many Miles of Driving Would It Take to Demonstrate Autonomous Vehicle Reliability ?" *Transportation Research Part A*, 2016, doi:10.1016/j.tra.2016.09.010.
4. National Highway Traffic Safety Administration, "Traffic Safety Facts 2010, a Compilation of Motor Vehicle Crash Data from the Fatality Analysis Reporting System and the General Estimates System," 2012, https://crashstats.nhtsa.dot.gov/Api/Public/ViewPublication/811659.
5. Ministry of Public Security Traffic Management Bureau, "Annual Report of the People's Republic of China on Road Traffic Accidents," 2014.
6. Najm, B.W G, Smith, J.D., and Yanagisawa, M., "Pre-Crash Scenario Typology for Crash Avoidance," *Dot Hs* (2010):767.
7. Kondoh, T. et al., "Identification of Visual Cues and Quantification of Drivers' Perception of Proximity Risk to the Lead Vehicle in Car-Following Situations," *Journal of Mechanical Systems for Transportation & Logistics* 1.2, no. 2008 (2008): 170-180.
8. Deery, H.A. and Fildes, B.N., "Young Novice Driver Subtypes: Relationship to High-Risk Behavior, Traffic Accident Record, and Simulator Driving Performance," *Human Factors* 41, no. 4 (1999): 628-643.
9. Dickie, D.A. and Boyle, L.N., "Drivers' Understanding of Adaptive Cruise Control Limitations," *Human Factors & Ergonomics Society Annual Meeting Proceedings* 53, no. 23 (2009): 1806-1810.
10. General Administration of Quality Supervision, Inspection and Quarantine of the People 's Republic of China, "GB 5768.3-2009 Road Traffic Signs and Markings Part 3 Road Traffic Markings," 2009.
11. General Administration of Quality Supervision, Inspection and Quarantine of the People 's Republic of China,"GB14886-2006 Setting and Installation Specifications of Road Traffic Signal," 2006.
12. General Administration of Quality Supervision, Inspection and Quarantine of the People 's Republic of China,"GB14887-2011 Road Traffic Signal Lights," 2011.
13. Fisher, R.A., "On the Interpretation of $\chi 2$ from Contingency Tables, and the Calculation of P," *Journal of the Royal Statistical Society* 85, no. 1 (1922): 87-94, JSTOR, JST0R, www.jstor.org/stable/2340521.
14. Strand, S., "Statistics for Research: With a Guide to SPSS (2nd ed.). George Argyrous, London: Sage Publications. 2005. Pbk £24.99 ISBN 1-4129-19487," *British Journal of Educational Psychology* 77, no. 2 (2007): 497.
15. TASS International, https://tass.plm.automation.siemens.com/prescan.

16. Zhao, D. et al., Accelerated Evaluation of Automated Vehicles Using Extracted Naturalistic Driving Data, *The Dynamics of Vehicles on Roads and Tracks* (2016).
17. Ljung Aust, M. and Engström, J., "A Conceptual Framework for Requirement Specification and Evaluation of Active Safety Functions," *Theoretical Issues in Ergonomics Science* 12, no. 1 (2011): 44-65.
18. INI Benchmark, http://benchmark.ini.rub.de/.

CHAPTER 3

Motion Planning of Vehicle Obstacle Avoidance in Complex Traffic Scenarios

Dongfang Dang

CHAPTER 3

With the increasing complexity, dynamicity and uncertainty of traffic, motion planning of automatic driving is getting more difficult and challenging. This paper focuses on the real-time motion planning problem of CAVs (connected and automated vehicles) in complex traffic scenarios. To effectively solve this problem, a general driving risk model is presented, which contains the following two essential parts: i) collision risk, i.e., the collision risk between the SV (subject vehicle) and other surrounding vehicles, pedestrians, buildings etc.; ii) non-collision risk, such as violation of traffic regulations, the deviation from the intention of driver, etc. To achieve the real time collision detection, the SV is approximated to a point and its shape is considered by extending the dimension of obstacles considering their relative position and velocity. Then an index similar to the exponential function is defined to calculate the collision risk value, which is composed of the collision detection result and TTC (time to collision). The index of non-collision risk is measured by the deviation from the desired states including regulations, driving manners and intentions etc. Accordingly, a motion planning algorithm by minimizing the aforementioned risk is proposed to calculate the required speed and yaw angle, considering the constraints of vehicle dynamics. Bench tests have been carried out to demonstrate the effectiveness of the proposed

CITATION: Dang, D., "Motion Planning of Vehicle Obstacle Avoidance in Complex Traffic Scenarios," SAE Technical Paper 2017-01-1958, 2017, doi:10.4271/2017-01-1958.

motion planning algorithm. The results show that it can safely handle a variety of complex traffic scenarios, such as lane change with a front vehicle cutting in suddenly, meanwhile the traffic regulations and vehicle dynamics constraints can also be met.

Introduction

Automated driving vehicles have demonstrated its potential in improving traffic safety and capacity [1, 2]. With the rapid development of sensors, artificial intelligence, computer and communication technology, automatic driving has been widely studied in both industrial and academic world [3]. The motion planning is an essential and fundamental component of automatic driving system [4]. It plans out a collision-free driving path or vehicle state, and controls the vehicle in real time.

The main motion planning methods can be roughly divided into three categories: search-based method [5, 6, 7], optimization control method [8, 9, 15, 16] and field-based method [10, 11, 12, 13, 14, 15, 16]. A* [5, 6] and Rapidly-exploring Random Trees [7] are common search-based methods. These methods are skilled in planning long-term collision-free trajectory and avoiding local minima. However, the planning trajectory is not optimal and the curvature is discontinuous. In addition, they may easily lead to stop-and-redirect planning since the velocity planning is not considered. Model predictive control is an effective optimization control method for motion planning. A scenario-based model predictive control method with two stages is proposed in [8]. The first stage determines the safe lane change planning based on traffic estimation and prediction via scenarios. The model predictive controller tracks the planning result in the second stage. In order to improve the robustness of the controller under disturbance and model uncertainty, a robust nonlinear model predictive frame work is presented in [9]. The potential field method is introduced by [10] and has been applied in the robotics field. It has the advantage of low computational complexity even in complex conditions with multiple obstacles. To plan the motion in lane keeping scenario, a hazard map was proposed in [11] based on potential fields. The concept of elastic bands used in robotics was extended to the vehicle motion planning, which significantly reduces the reactive behavior of the lane-keeping system. A driving safety field model based on field theory is proposed in [13]. The model considers the interactions between driver, vehicle and road. An application of this model in pre-collision warning is introduced in [14]. There has been substantial research on path planning and tracking for collision avoidance systems. However, the vehicle stability and the path traceability may not be guaranteed since the constraints imposed by vehicle dynamics is not taken into consideration. Ref. [15] improved the path tracking ability of vehicle by constructing a potential field that considers the constraints of road and obstacles. Furthermore, [16] proposed a way to combine the potential field method and model predictive control to utilize the advantages of these two methods.

The driving safety is improved by motion planning methods significantly. However, there are some drawbacks to the existing methods. Firstly, the aforementioned methods may find out a wrong path since the shapes of the obstacles are not considered explicitly. Secondly, most approaches are merely suitable in simple traffic scenarios. Nowadays, sufficient information is available due to the development of sensor and communication technologies, which requires the algorithm to deal with more complex conditions.

To overcome these drawbacks, this paper presents a motion planning method for CAVs based on general driving risk model. The proposed model consists of collision risk and non-collision risk. To evaluate the collision risk between SV and obstacles, a real-time collision detection algorithm is presented. The shapes of the SV and obstacles can be considered explicitly. In addition, the constraints of road, traffic regulations and intention of driver are considered in non-collision risk model. Moreover, a minimum risk search algorithm is proposed to generate the collision-free state of vehicle at every step size. And the lateral and longitudinal controllers are used to realize the desired vehicle states.

System Architecture

The overall architecture of the obstacle avoidance system is shown in Figure 1, which is composed of four modules: road & obstacles information, motion planning, state tracking and vehicle dynamics. Radars, cameras, etc. are used in the perception module [17, 19]. With the development of wireless communication, traffic participates can collect and share information by V2X (vehicle-to-X) technologies [18]. A low-layer controller is used to control the steering and speed of CAV [7, 19]. The information of road, velocity (v_i), yaw (ψ_i) and position (X_{Vi} Y_{Vi}) of obstacles is obtained by the road & obstacles information module. The desired state of SV is generated by motion planning module considering constraints of vehicle dynamics. The desired state includes velocity (v_{des}) and yaw (ψ_{des}). Moreover, the lateral and longitudinal controllers aim to track the desired state of SV. The output of lateral controller is the steering angle command (δ_{SW}). The longitudinal controller outputs two control signals, i.e., brake (P_b) and throttle pedal (P_t). This paper mainly focuses on the motion planning module, which is critical to automatic driving.

FIGURE 1 Architecture of the obstacle avoidance system.

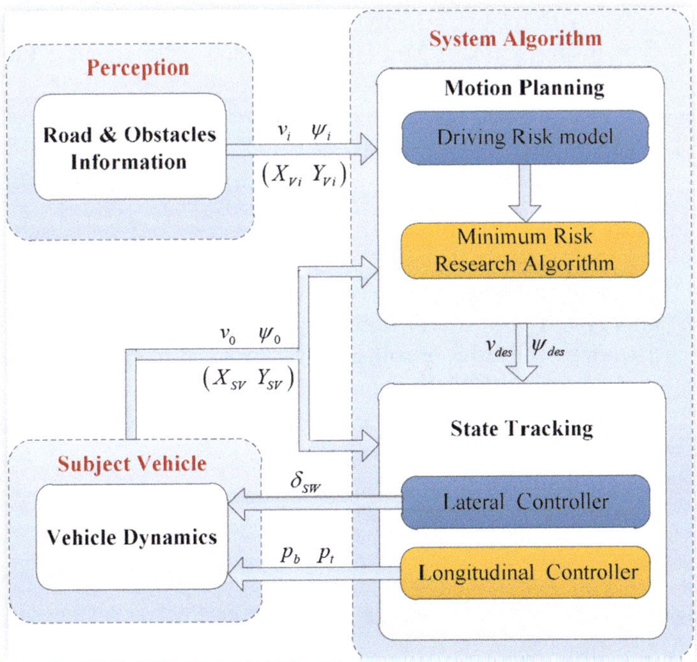

General Driving Risk Model

The general driving risk model proposed in this paper takes road, obstacles, traffic rules and driver intention into consideration comprehensively. It is expressed as:

(1)

where α_c, α_{ld}, α_{cd}, α_s and α_d are weight factors of corresponding risk model, the subscript i represents the i-th obstacle and subscript 0 represents SV, R_{ci} is the collision risk, R_{cd} and R_{ld} are the risk of deviation from road boundary and centerline respectively, R_s is the speeding risk, R_d is the risk of deviation from the intention of driver, v_0 and ψ_0 are velocity and yaw respectively.

Collision Risk Model

With the increasing complexity of traffic scenarios, the shape and velocity direction must be considered explicitly in obstacle avoidance system, especially with the increasing number of obstacles. An expanded circle based on the geometry of SV and obstacle is used in [20, 21, 22] to planning the motion. However, this leads to more conservativeness because most obstacles are rectangle. Here an expanded ellipse is proposed to deal with the non-uniformity of shape.

REAL-TIME COLLISION DETECTION

The velocity obstacle method was first introduced by Fiorini et al. [20] and has been widely applied in obstacle avoidance system. Let A and B denotes the SV and obstacle respectively. Object A can be reduced to a point by growing B to an ellipse with the dimension of A as shown in Figure 2 [23].

FIGURE 2 The paradigm of velocity obstacle.

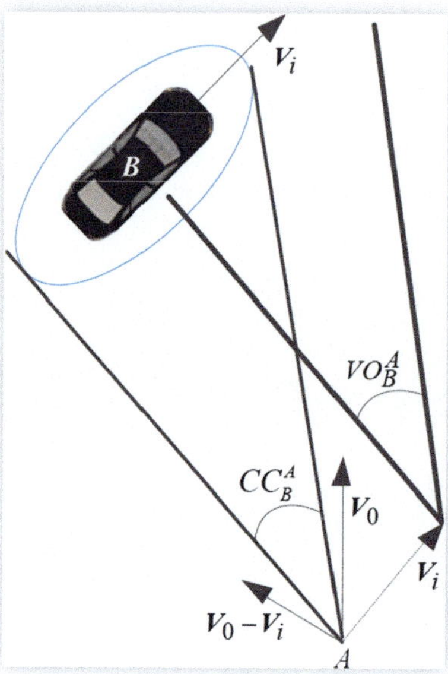

In Figure 2, $\boldsymbol{V}_0$ and $\boldsymbol{V}_i$ denotes the velocity of SV and i-th obstacle. To detect the collision more simply, B is assumed to be stationary and SV moves with relative velocity $\boldsymbol{V}_0$ - $\boldsymbol{V}_i$. Then a collision may occur if $\boldsymbol{V}_0$ - $\boldsymbol{V}_i$ remains in CCBA$^{CC_B^A}$. Similarly, a collision will happen if $\boldsymbol{V}_0$ remains in VOBA$^{VO_B^A}$. The collision detection results is

(2)

COLLISION RISK

In reality, a higher relative velocity, shorter distance and more kinetic energy between SV and obstacles indicate a greater collision risk. Therefore, the index of collision risk is determined considering the collision injury and TTC. The conservation laws of momentum (3) and energy (4) are used to compute the velocity of SV (v_0') and obstacle (v_i') after the collision:

(3)

(4)

where v_0 and m_0 represent the velocity and mass of SV, m_i is the mass of i-th obstacle, ΔE is the total loss of kinetic energy. Combining (3) and (4) yields:

(5)

In order to get the maximum loss of kinetic energy, the differential of ΔE is

(6)

Let $\mathrm{d}(\Delta \mathrm{E})\mathrm{d}(\mathrm{v}0') = 0$ $\frac{d(\Delta E)}{d(v_0')}=0$ and we obtain the following condition that the kinetic energy loss is maximum:

(7)

The objects are different, the damage caused by the same kinetic energy is also different. For example, the same kinetic energy generates more harm to humans than animals. And so injury factors η_0 and η_i are used to describe the damage property of SV and obstacles. The injury I_i between SV and the i-th obstacle is

(8)

Based I_i and TTC, the collision risk index is calculated by

(9)

(10)

where R_c^* is normal risk index in typical collision scenario, TTC_{thr} is the threshold of TTC, S_i is the distance between SV and the i-th obstacle. Figure 3 sketches a traffic scenario with multiple obstacles, where V1, ⋯ , V4 are obstacles and v_1, ⋯ , v_4 are their velocity. The collision risk index is shown in Figures 4 and 5.

A similar risk index is defined as the following to represent the risk of lane departure:

(11)

where R_{ld}^* is the normalization parameter, TLC_{thr} (Time to Line Crossing) is the threshold of time to lane crossing, I_l is the collision injury between SV and infrastructures.

FIGURE 3 Position and state about SV and obstacles.

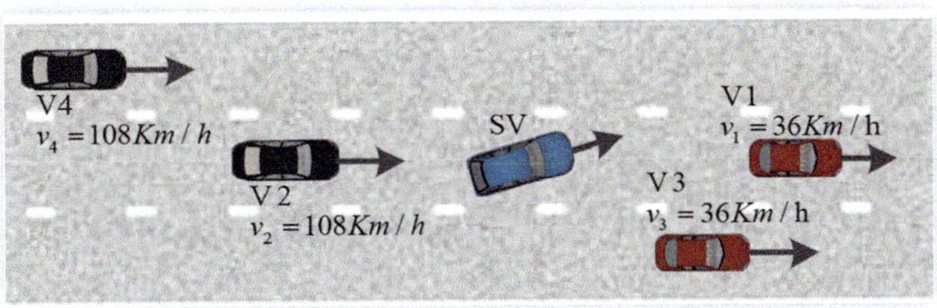

FIGURE 4 Risk index between SV and V1, V2.

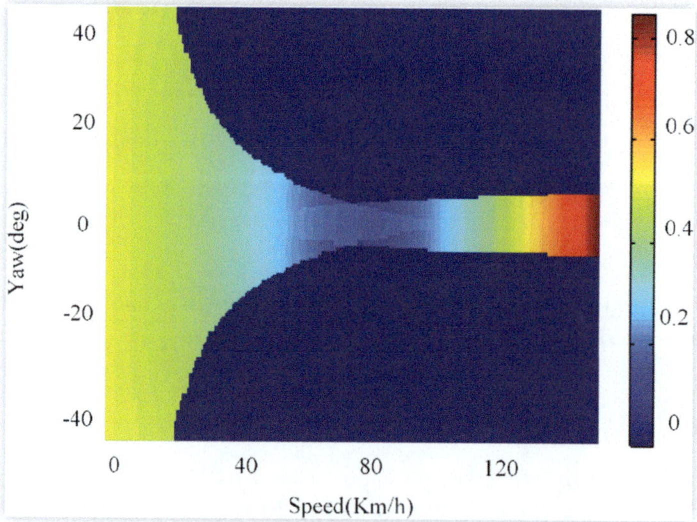

FIGURE 5 Risk index between SV and V3, V4.

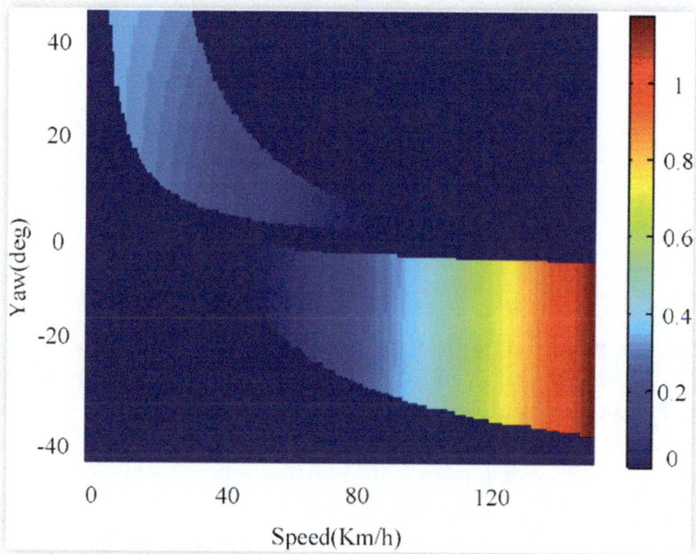

Non-Collision Risk Model

In addition to the main task of obstacle avoidance, the motion planning should observe traffic regulations and meet the driver's intention. Therefore, non-collision risk is proposed to evaluate the deviation from traffic regulations and driver's intention.

VIOLATING TRAFFIC REGULATIONS RISK

Centerline Departure Risk. The traffic regulation requires driver not to change lanes casually. Therefore, SV should drive along the lane centerline as close as possible at normal conditions. The centerline departure risk is defined according to the deviation from the centerline:

(12)

where y_{ll} is the distance between the c.g.(center of gravity) of SV and the right lane line, W_l is the width of lane, $\Delta\psi$ is the yaw angle of SV, R^*_{cd} is the normalization parameter.

Over Speed Risk. Drivers are prohibited from over speeding. Therefore, over speed risk R_s is defined as

(13)

where R_s^* is the normalization parameter, v_{lmi} is the velocity limit.

Risk of Deviation from Driver's Intention. When the vehicle's operation state does not meet the intention of driver, it can affect the driver's psychology and even the driving safety. And so the risk of deviation from driver's intention is modeled as follow:

(14)

where R_d^* is the normalization parameter, v_d is driver's desired velocity.

Motion Planning and Control

Minimum Risk Search Algorithm

To guarantee the stability of vehicle dynamics, the proposed motion planning algorithm takes into consideration the maximum acceleration and yaw rate of SV. The dynamic constraints are

(15)

(16)

where Δt is the step size, a_{dmax}, a_{amax}, ω_{max}, a_{yup}, v_{0max} are the maximum deceleration, acceleration, yaw rate, lateral acceleration, velocity of SV respectively. g is the gravitational acceleration. Considering the driving comfort, $a_{yup} = 0.4\ m/s^2$. Then the constraints are expressed as:

(17)

The minimum risk search algorithm is described in Algorithm 1.

FIGURE 6 Vehicle bicycle model.

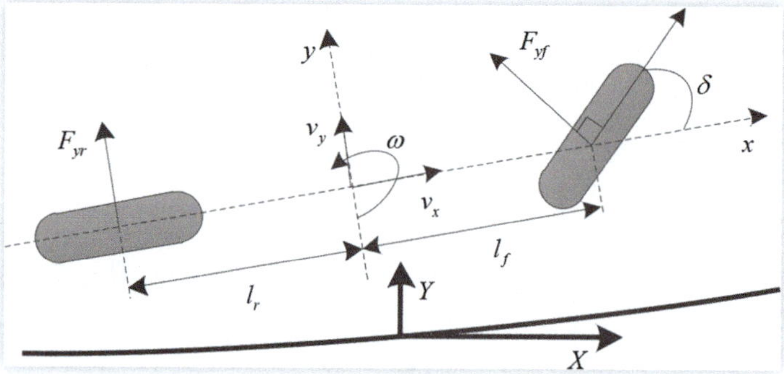

ALGORITHM 1. Minimum Risk Search Algorithm

Input: v_i, ψ_i, v_0, ψ_0
Parameters: $l_i, w_i, l_0, w_0, v_{max}, \psi_{max}$
Output: v_{des}, ψ_{des}

for $i = 1:3$
    $v_c = v_0 - v_{max} + (i-1)v_{max}$
    for $j = 1:5$
        $\psi_c = \psi_0 - \psi_{max} + (j-1)\psi_{max}/2$
        $R_c(i,j) = R(v_i, \psi_i, v_c, \psi_c)$
    end
end
$[m,n] = find(R_c == min(min(R_c)))$
$v_{des} = v_0 - v_{max} + (m-1)v_{max}$
$\psi_{des} = \psi_0 - \psi_{max} + (n-1)\psi_{max}/2$

Vehicle Dynamics and State Tracking

The dynamics of SV is modeled as a bicycle model as shown in Figure 6 [9, 16].

The dynamic equations are:

 (18)

where v_x and v_y denote the longitudinal and lateral velocity of SV, $\Delta\psi$ is the yaw error, ω and I_z denote the yaw rate and yaw inertia, F_x is the total longitudinal force of tires, F_{yf} and F_{yr} are the total lateral forces of the front and rear tires, l_f and l_r are the distances of the front tire and the rear tire respectively from the c.g. of the vehicle. For small slip-angles, lateral tire force is proportional to the slip-angle [24]:

 (19)

where $C_{\alpha f}$ and $C_{\alpha r}$ the cornering stiffness of the front and rear tire, α_f and α_r are the sideslip angles of the front and rear tires, δ is the steering angle. The detailed information about the tracking controllers are refer to [7, 19].

Experimental Results and Discussions

To verify the effectiveness of the proposed approach, bench tests are conducted in a series of complex traffic scenarios. The block diagram of the bench system is shown in

FIGURE 7 The bench testing system.

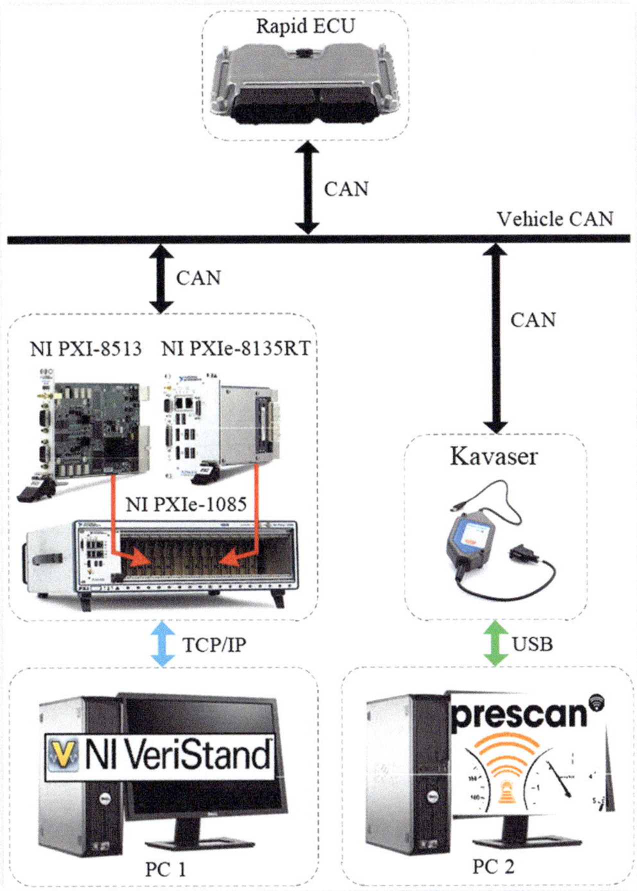

Figure 7. The scenarios are generated in PreScan. The motion planning and tracking algorithm are implemented on a Rapid Prototyping ECU. The vehicle dynamical models run in the NI PXI platform.

Experimental Scenarios

The speed limit for all experimental scenarios is 100 Km/h. The bench tests are conducted under the following scenarios:

- **Scenario 1**: V1, ⋯, V5 are obstacles around SV. The initial states are shown in Figure 8. The obstacles travel straightly, the driver's desired velocity is 90 Km/h and $\omega_{max} \geq 9.1$ *deg/s* when $v_0 \leq 90$ *Km/h*.
- **Scenario 2**: V1 and V2 are obstacles and the initial states are shown in Figure 10. The radius of curve road is 250 m. V1 driving along the current lane, V2 changes its lane to right aggressively at X = 60 m. The driver's desired velocity is 72 Km/h and $\omega_{max} \geq 11.4$ *deg/s* when $v_0 \leq 72$ *Km/h*.
- **Scenario 3**: V1 and V2 are obstacles and the initial stats are shown in Figure 12. V1 changes its lane to left aggressively at X = 240 m, and V2 change its lane to right aggressively at X = 110 m. The driver's desired velocity is 90 Km/h and $\omega_{max} \geq 9.1$ *deg/s* when $v_0 \leq 90$ *Km/h*

FIGURE 8 Initial states of SV and obstacles under **Scenario 1**.

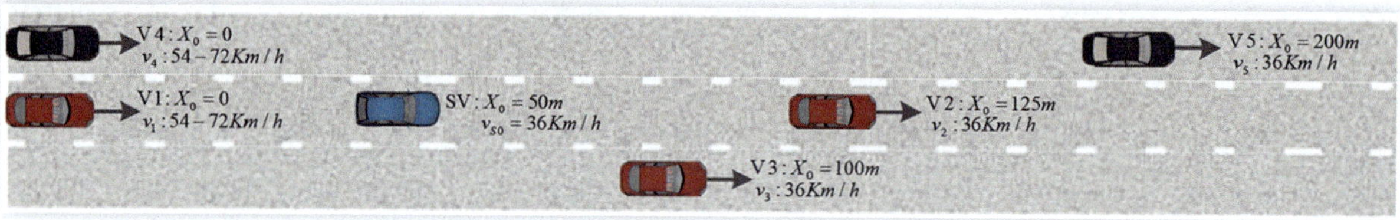

FIGURE 9 Results of **Scenario 1**. (a) Paths of SV and obstacles; (b) Velocity and yaw error of SV; (c) Acceleration and yaw rate of SV.

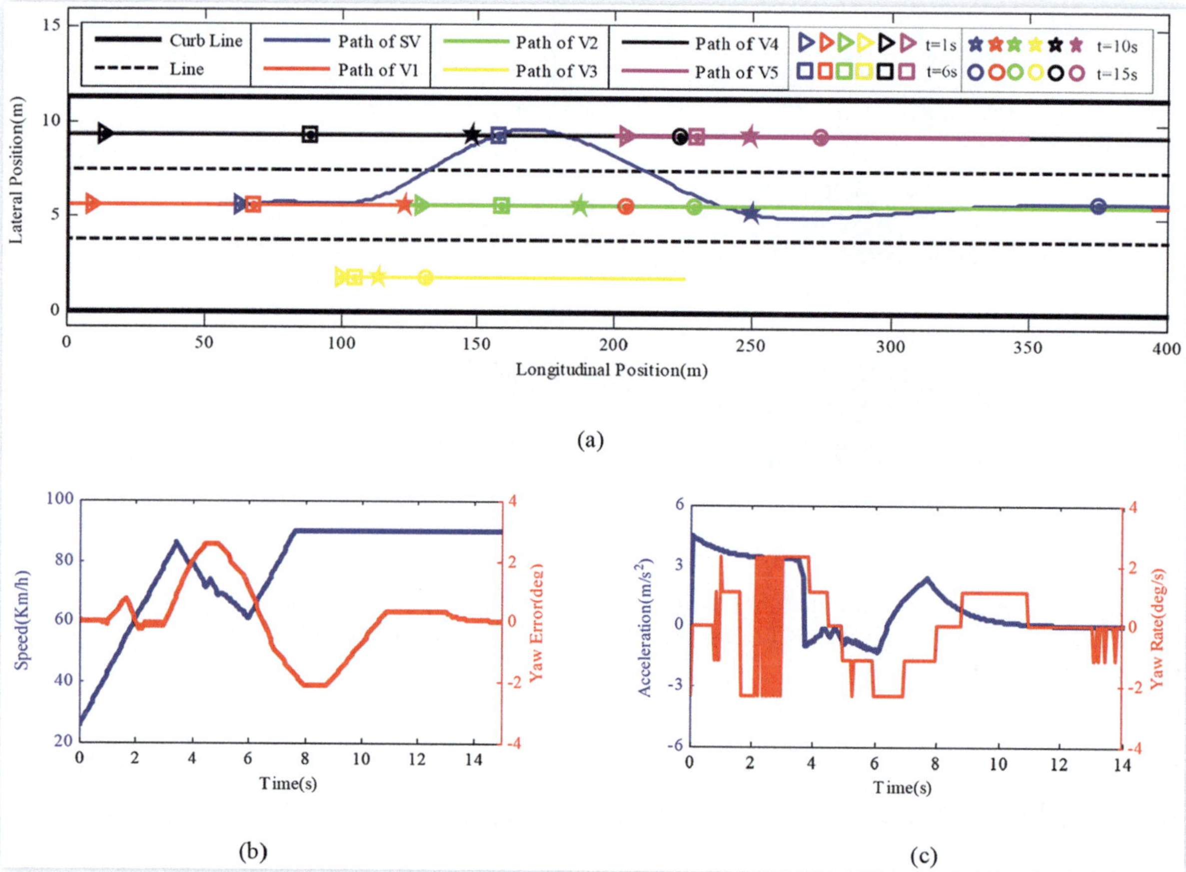

FIGURE 10 Initial states of SV and obstacles under **Scenario 2**.

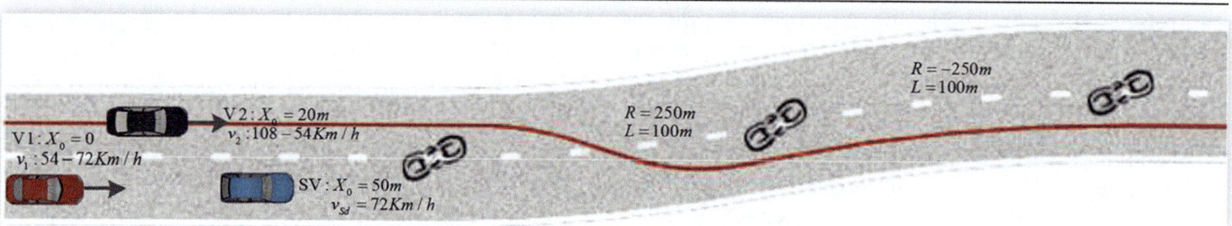

FIGURE 11 Results of **Scenario 2**. (a) Paths of SV and obstacles; (b) Velocity and yaw error of SV; (c) Acceleration and yaw rate of SV.

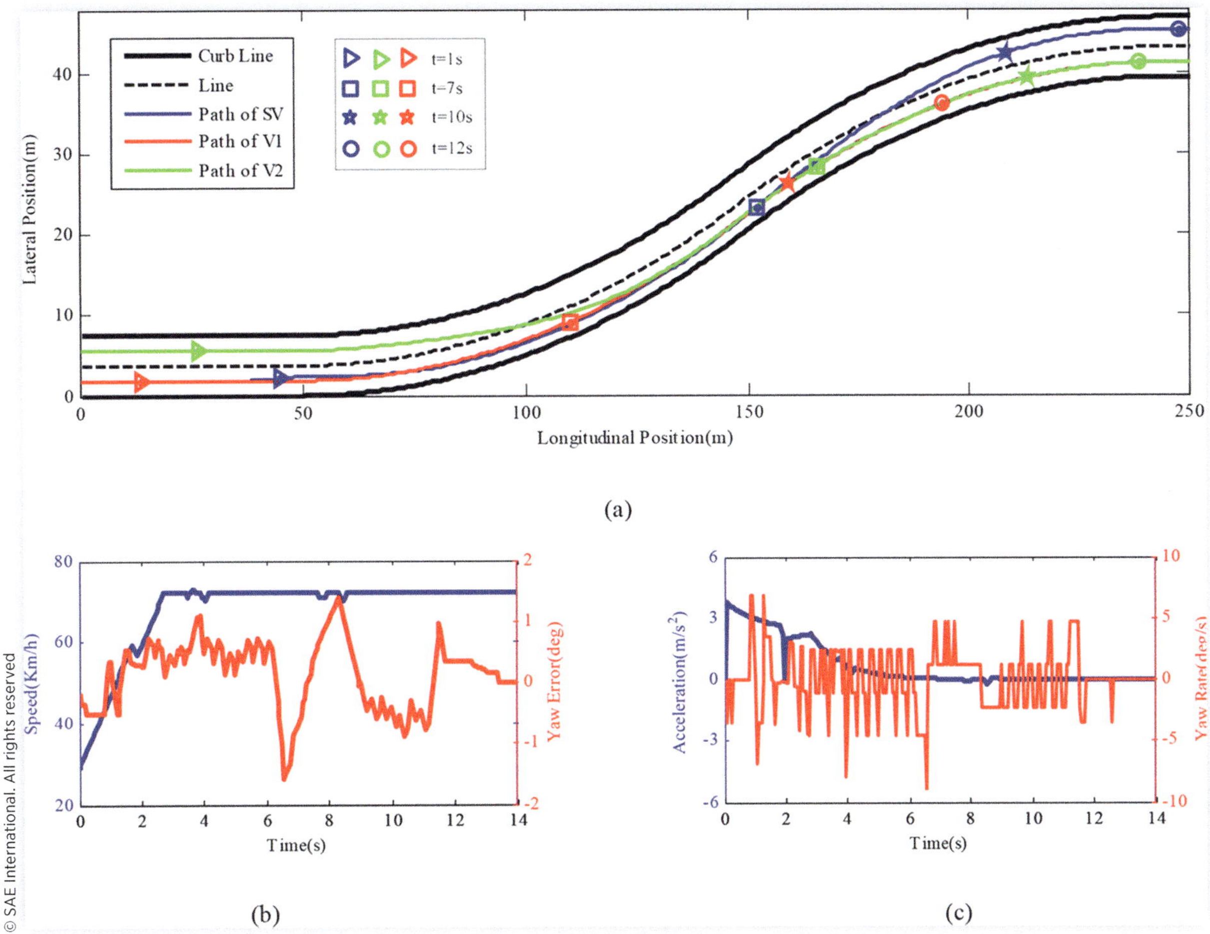

FIGURE 12 Initial states of SV and obstacles under **Scenario 3**.

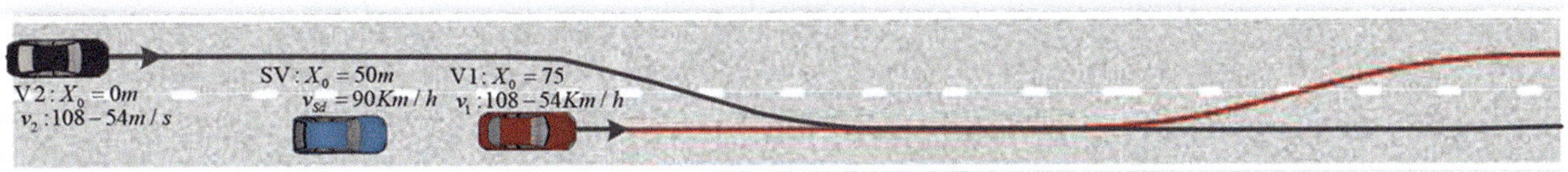

Scenario 1 is mainly designed to verify the obstacle avoidance ability in a narrow motion planning space. Under **Scenario 2**, vehicles run on a curve road with an obstacle cutting in aggressively. **Scenario 3** is to verify the motion planning algorithm in a more serious condition that obstacles cutting in continuously.

Results and Analysis

The simulation results of **Scenario 1** are shown in Figure 9. In Figure 9(a), each color represents an obstacle and each shape represents different time. To avoid V1, V2, V3 and V4, SV makes the decision to turn left with accelerating firstly. Then, SV turns right to

FIGURE 13 Results of **Scenario 3**. (a) Paths of SV and obstacles; (b) Velocity and yaw error of SV; (c) Acceleration and yaw rate of SV.

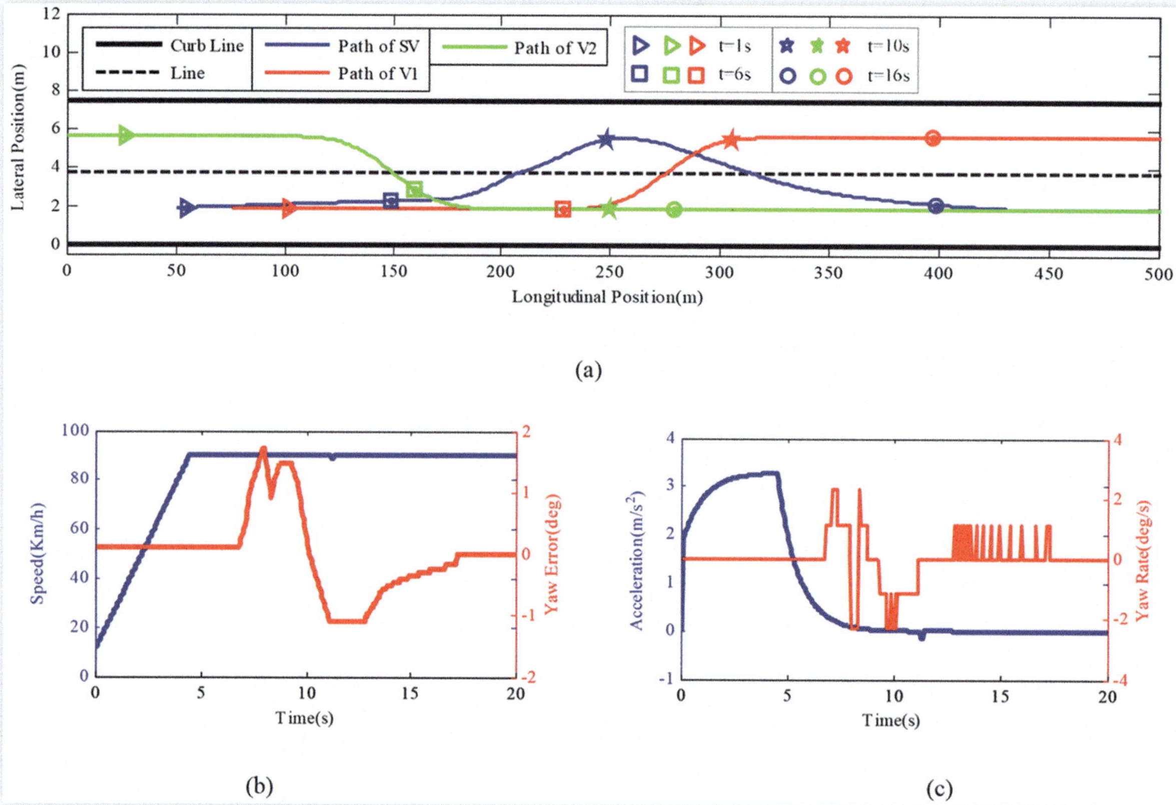

avoid V5 and goes on travelling along the centerline. From Figure 9(b), SV travels with the desired velocity and the control error converges zero. From Figure 9(c) the acceleration and yaw rate are within the required range in the whole motion planning process. Simulation results of **Scenario 2** and **3** are shown in Figure 11 and Figure 13 respectively. All the results under different scenarios show that the proposed motion planning algorithm has a good performance in complex and dynamic traffic scenarios.

Conclusions

Motion planning in complex and dynamic traffic scenarios remains a challenging research problem. In this paper, a motion planning method for CAVs is presented. Bench experiments show that the SV controlled by the proposed algorithm can avoid the obstacles successfully in a variety of complex traffic scenarios. Meanwhile, the constraints of road, traffic regulations and intention of driver are also met as required by considering a non-collision risk. In the next, the proposed motion planning method may be improved by: (1)

Considering the vehicle dynamical properties when planning the motion; (2) The uncertainty of road users may be considered by using a probabilistic risk model; (3) Further tests should be carried out under more complex scenarios.

Contact Information

Feng Gao
School of Electrical Engineering, Chongqing University,
Chongqing, China
gaofeng1@cqu.edu.cn

DongFang Dang
School of Electrical Engineering, Chongqing University,
Chongqing, China
sunshinedang@163.com

Tao Chen
China Automotive Engineering Research Institute Co., Ltd,
Chongqing, China
chentao@caeri.com.cn

Cai Yang
Changan Automobile New Energy Research Institute,
Chongqing, China
yangcai@changan.com.cn

Acknowledgments

This work was supported by National key research and development project under grant 2016YFB0101104 and 2016YFB0100900, Scientific Technological Plans of Chongqing under Grant cstc2015zdcy-ztzx60002 and Industrial Base Enhancement Project under grant 0714-EMTC02-5593/20.

Abbreviations

CAV - connected and automated vehicle
SV - subject vehicle
TTC - time to collision
V2X - vehicle-to-X
TLC - time to lane crossing
c.g. - center of gravity
CAN - Control Area Network

References

1. Wang, J., Steiber, J., and Surampudi, B., "Autonomous Ground Vehicle Control System for High-Speed and Safe Operation," *American Control Conference 2008*, USA, June 11-13, 2008, doi:10.1109/ACC.2008.4586494.
2. Rios-Torres, J. and Malikopoulos, A., "A Survey on the Coordination of Connected and Automated Vehicles at Intersections and Merging at Highway on-Ramps," *IEEE Transactions on Intelligent Transportation Systems* 18, no. 5 (2017): 1066-1077, doi:10.1109/TITS.2016.2600504.

CHAPTER 3

3. Li, X., Sun, Z., Cao, D., He, Z., and Zhu. Q., "Real-Time Trajectory Planning for Autonomous Urban Driving: Framework, Algorithms, and Verifications," *IEEE/ASME Transactions on Mechatronics* 21, no. 2 (2016): 740-753, doi:10.1109/TMECH.2015.2493980.

4. Tanzmeister, G., Friedl, M., Wollherr, D., and Buss. M., "Efficient Evaluation of Collisions and Costs on Grid Maps for Autonomous Vehicle Motion Planning," *IEEE Transactions on Intelligent Transportation Systems* 15, no. 5 (2014): 2249-2260, doi:10.1109/TITS.2014.2313562.

5. Dolgov, D., Thrun, S., Montemerlo, M., and Diebel, J., "Path Planning for Autonomous Vehicles in Unknown Semi-Structured Environments," *International Journal of Robotics Research* 29, no. 5 (2010): 485-501, doi:10.1177/0278364909359210.

6. Fassbender, D., Mueller, A., and Wuensche, H., "Trajectory Planning for Car-Like Robots in Unknown, Unstructured Environments," *IEEE/RSJ International Conference on Intelligent Robots & Systems*, USA, September 14-18, 2014, doi:10.1109/IROS.2014.6943071.

7. Kuwata, Y., Teo, J., Fiore, G., Karaman, S. et al., "Real-Time Motion Planning with Applications to Autonomous Urban Driving," *IEEE Transactions on Control Systems Technology* 17, no. 5 (2009): 1105-1118, doi:10.1109/TCST.2008.2012116.

8. Schildbach, G. and Borrelli, F., "Scenario Model Predictive Control for Lane Change Assistance on Highways," *IEEE Intelligent Vehicles Symposium*, South Korea, June 28-July 1, 2015, doi:10.1109/IVS.2015.7225752.

9. Gao, Y., Gray, A., and Tseng, H., "A Tube-Based Robust Nonlinear Predictive Control Approach to Semiautonomous Ground Vehicles," *Vehicle System Dynamics* 52, no. 6 (2014): 802-823, doi:10.1080/00423114.2014.902537.

10. Krogh, B., "A Generalized Potential Field Approach to Obstacle Avoidance Control," *SME Conference Proceeding Robotics Research*, USA, 1984.

11. Thomas, S. and Thorsten, B., "From Robotics to Automotive: Lane-Keeping and Collision Avoidance Based on Elastic Bands," *Vehicle System Dynamics* 46, no. 7 (2008): 597-619, doi:10.1080/00423110701543452.

12. Wolf, M. and Burdick, J., "Artificial Potential Functions for Highway Driving with Collision Avoidance," *IEEE International Conference on Robotics & Automation*, USA, May 19-23, 2008, doi:10.1109/ROBOT.2008.4543783.

13. Wang, J., Wu, J. and Li, Y., "The Driving Safety Field Based on Driver-Vehicle-Road Interactions," *IEEE Transactions on Intelligent Transportation Systems* 16, no. 4 (2015): 2203-2214, doi:10.1109/TITS.2015.2401837.

14. Wang, J., Wu, J., Zheng, X., Ni, D. et al., "Driving Safety Field Theory Modeling and Its Application in Pre-Collision Warning System," *Transportation Research Part C Emerging Technologies* 72 (2016): 306-324, doi:10.1016/j.trc.2016.10.003.

15. Ji, J., Khajepour, A., Melek, W., and Huang, Y., "Path Planning and Tracking for Vehicle Collision Avoidance Based on Model Predictive Control with Multi-Constraints," *IEEE Transactions on Vehicular Technology* 66, no. 2 (2017): 952-964, doi:10.1109/TVT.2016.2555853.

16. Rasekhipour, Y., Khajepour, A., Chen, S. K., and Litkouhi, B., "A Potential Field-Based Model Predictive Path-Planning Controller for Autonomous Road Vehicles," *IEEE Transactions on Intelligent Transportation Systems* 18, no. 5 (2017): 1255-1267, doi:10.1109/TITS.2016.2604240.

17. Chu, K., Lee, M., and Sunwoo, M., "Local Path Planning for Off-Road Autonomous Driving with Avoidance of Static Obstacles," *IEEE Transactions on Intelligent Transportation Systems* 13, no. 4 (2012): 1599-1616, doi:10.1109/TITS.2012.2198214.

18. Seredynski, M. and Viti, F., "A Survey of Cooperative ITS for Next Generation Public Transport Systems," *2016 IEEE 19th International Conference on Intelligent Transportation Systems*, Brazil, November 1-4, 2016, doi:10.1109/ITSC.2016.7795714.

19. Ma, L., Xue, J., Kawabata, K., Zhu, J. et al., "Efficient Sampling-Based Motion Planning for On-Road Autonomous Driving," *IEEE Transactions on Intelligent Transportation Systems* 16, no. 4 (2015):1961-1976, doi:10.1109/TITS.2015.2389215.

20. Fiorini, P. and Shiller, Z., "Motion Planning in Dynamic Environments Using the Relative Velocity Paradigm," *IEEE International Conference on Robotics and Automation*, USA, May 2-6, 1993, doi:10.1109/ROBOT.1993.292038.

21. Choi, J., "Real-Time Obstacle Avoiding Motion Planning for Autonomous Ground Vehicles," Ph.D. thesis, Computer Engineering Department, University of California, Santa Cruz, 2010.

22. Elbanhawi, M., Simic, M., and Jazar, R., "Randomized Bidirectional B-Spline Parameterization Motion Planning," *IEEE Transactions on Intelligent Transportation Systems* 17, no. 2 (2016):406-419, doi:10.1109/TITS.2015.2477355.

23. Latombe, J., "Robot Motion Planning, second Edition," *New York, Communications on Pure & Applied Mathematics* (1991): 105-120, doi:10.1007/978-1-4615-4022-9.

24. Rajamani, R., *Vehicle Dynamics and Control*, 2nd ed. (New York, Springer, 2006), 15-30, doi:10.1007/978-1-4614-1433-9.

CHAPTER 4

Driver Behavior While Operating Partially Automated Systems: Tesla Autopilot Case Study

John Shutko and Benjamin Osafo-Yeboah
Ford Motor Co., Ltd.

Chris Rockwell and Mark Palmer
Lextant

Level 2 (L2) partially automated vehicle systems require the driver to continuously monitor the driving environment and be prepared to take control immediately if necessary. One of the main challenges facing developers of these systems is how to ensure that drivers understand their role and stay alert as the systems require. With little real world data, it has been difficult to understand user attitudes and behaviors toward the implementation and use of partially automated vehicles. At the time of this study, Tesla was one of the few OEMs with a partially automated vehicle feature available on the market; Autopilot. In order to understand how customers interact with a partially automated vehicle, a study was conducted to observe people driving their own Tesla vehicles while autopilot was engaged.

Sixteen Tesla owners (14 males and 2 females) between ages 25 to 60 had their vehicles instrumented with video/audio data collection systems for three consecutive days. These owners were dedicated autopilot users who used the feature daily and primarily on highways. Results from the study show that (i) participants' eye off-road glance behavior while operating a partially automated vehicle was somewhat similar to eye off-road glance behavior in radio tuning task while driving manually, though substantially more variable, (ii) eyes on/off road

CITATION: Shutko and J., Osafo-Yeboah, B., Rockwell, C., and Palmer, M., "Driver Behavior While Operating Partially Automated Systems: Tesla Autopilot Case Study," SAE Technical Paper 2018-01-0497, 2018, doi:10.4271/2018-01-0497.

glance behavior had no relation to whether drivers kept their hands on the steering wheel or not, (iii) drivers exhibit a bi-modal behavior when operating a partially automated vehicle: those who mostly kept their hands on steering wheel while autopilot was active (i.e. *active drivers*) verses those who mostly kept their hands off the steering wheel while autopilot was active (i.e. *supervisors*).

Introduction

Automobiles have come a long way since they were introduced over a century ago. In the United States today, there are more than one car per licensed driver. This is a reflection of the need for individuals to drive in order to participate effectively in the modern economy, and to maintain an enriching social life [1]. While driving allows economic participation and social enrichment, it is also a leading cause of death in the United States. Over thirty thousand lives are lost each year on American roads, mostly attributed to human error [2]. Recent advances in technology has led to the development and deployment of advanced driver assistance systems (ADAS) with the goal of enhancing safe and efficient driving. While ADAS systems enhance safe driving, they also serve as building blocks for the eventual development of automated vehicles. Currently, SAE defines six levels of automated driving systems: level 0 through level 5, with level 0 having no automation, while level 5 is fully autonomous driving.

Partially automated driving systems, called Level 2 (L2) systems, are defined by SAE J3016 as automated vehicles in which the system performs the execution of steering and acceleration or deceleration of the vehicle while the human driver monitors the driving environment [3]. In other words, the driver must be alert and attentive, in order to detect and respond appropriately to events that are beyond the capabilities of the partially automated driving system. Therefore, in order for a partially automated vehicle to operate effectively, the driver must be alert at all times. However, it has been found that drivers' ability to perceive critical factors within their environment or to detect system changes or failures may be diminished relative to engaged and alert manual driving as their role shifts from active vehicle control to passive monitoring, primarily due to loss of awareness of the state and the processes of the system [4, 5].

One of the many challenges that developers of partially automated driving vehicles are grappling with currently is how to ensure that drivers understand their role and remain alert while operating a partially automated vehicle. In particular, there are concerns that drivers' willingness to engage in non-driving related tasks will be impacted while operating a partially automated vehicle due to the increased use of nomadic electronic devices, and the reduction in workload provided by these systems [6, 7]. However, little research has been done on driver attitude and behavior while operating a partially automated vehicle.

With much speculation and little experience, it has been difficult to forecast user attitudes and behaviors while using autonomous driving features. Currently Tesla is one of the few manufacturers who has a vehicle on the market with partially automated driving functionality available to the driver (i.e. autopilot). The goal of present study was to understand Tesla owners' driving behavior while autopilot is active, and to help understand what influence drivers' use of the feature. To help evaluate these goals, a naturalistic study was conducted using instrumented Tesla vehicles to collect data over a three-day period.

Methodology

Participants

Sixteen (16) drivers, between 25 to 60 years of age with autopilot version 8.0 equipped Tesla Model S vehicles were recruited from Phoenix, AZ and Columbus, OH metro areas through newspaper ads to participate in this study. Participants consisted of 14 males and 2 females, each with a valid drivers' license. Six (6) participants were considered novices (they had 4 months or less experience using autopilot), while 10 were considered experienced (they had more than 4 months experience using autopilot). Each participant had their Tesla vehicle outfitted with two discreet video/audio capture system that recorded all their drives over a 3-day period. Each participant received a $250 compensation for their participation.

Study Design

The study consisted of a 3-pronged approach of journaling, observation and interview. This approach allowed the authors to identify situations in which Tesla owners use autopilot, what they do when autopilot is enabled, and how they feel about using autopilot.

JOURNALING

One week prior to the observation and interview, participants created a journal of how often they used autopilot, what situations prompted them to use autopilot, and how they felt about their experience with autopilot.

OBSERVATION

Each participant had his/her Tesla Model S outfitted with two discreet video/audio capture systems that recorded all their drives over a 3-day period.

INTERVIEW

At the end of the observational period, the experimenters conducted a one hour retrospective interview to review, discuss, and understand participants' behaviors during the use of autopilot.

Procedure

One week before the observation and interview sections, participants documented their autopilot usage. Specifically, how often they used autopilot, the situation that prompted autopilot use, and the experience they had using autopilot.

After one week of journaling, participants drove their vehicles to a research facility where technicians outfitted their vehicle with audio/video capture system. Two audio/video recording systems were discreetly installed. A forward over the shoulder camera recorded the road scene and the cluster human machine interface (HMI), while a rear facing camera recorded both the driver's face and hands. The audio/video capture system was set up such that, the driver had to press a button to start the recording on each drive. Once instrumented, participants drove their vehicles as they typically would. The video/audio recording system, recorded all their drives for the next three days. No participant reported that they forgot to activate the audio/video system on any drive.

On the third day of observation, two experimenters met each participant at a prearranged location, and followed the participant during one of their typical drives.

During this drive, the experimenters used a remote video "beaming" system to remotely access and observe participants while they go about their typical errands. At the end of the trip, the experimenters conducted a one-hour interview with the participant. The interview included follow-up questions from the interactions observed during the remote video "beaming" session. The interview session was recorded for future analysis. The video/audio capture system was removed from participants' vehicle at the end of the third day.

Results

Qualitative Data

AUTOPILOT UNDERSTANDING AND USAGE SITUATIONS

All participants (16 out of 16) treated the autopilot feature as a driver assistance feature, and understood that the driver was primarily responsible for the safe operation of the vehicle even when autopilot was active. Participants used autopilot in a variety of situations both on and off highways. In particular, all participants cited one or both of the following as a reason for activating autopilot.

i. When lane markings were clearly visible, because they believed autopilot could track and keep them in the lane more effectively
ii. When they were in traffic situations, because they knew from experience that autopilot could track effectively the vehicle in front of their vehicle.

SYSTEM EXPECTATIONS

In a subjective evaluation, participants reported realistic expectations for Tesla's autopilot feature prior to purchase. No participant reported that he or she expected autopilot to drive for them without any driver monitoring. Instead, participants described autopilot as a driver assistance or backup feature. All 16 participants reported that they had completed extensive research (through forums, Tesla news updates, news articles, Tesla sales, and tech crew) on the capabilities and limitations of autopilot prior to purchase.

INITIAL EXPOSURE

Despite having realistic expectations prior to purchase, participants' expectations of autopilot dropped further during the first few months of ownership, as they experienced real world limitations of the feature. However, expectations improved after the initial few months of ownership as drivers learned how the system worked, and how and when it could be used. All 12 participants who had more than two months of experience with the system reported having less dissatisfaction and uncertainty in autopilot's capabilities after becoming familiar and comfortable with the system through positive driving experiences.

EFFECT OF AUTOPILOT ACCIDENT ON DRIVER BEHAVIOR

The fatal crash that killed a Tesla driver while using autopilot happened three months prior to this study, therefore, participants were asked if the crash had any impact on their use or attitude towards autopilot. All 16 participants reported that they considered the accident a "wake up call" and that it made them more conservative in their expectations of autopilot. Specifically, most participants reported that they changed their driving behavior by spending less time doing non-driving related tasks, and more time monitoring the road ahead when autopilot was active. In spite of the behavior change, participants were not deterred from using autopilot and unanimously attributed the accident to misuse of the feature.

Behavior Analysis

A detailed analysis of driver behavior while autopilot is engaged, was conducted using eye glance data and driver hands-on/off steering wheel data. The following two sections provide details on eye glance and hands-on steering wheel results.

Eye Glance Analysis

Overall, a total of 2210 eyes off-road glances (glances away from road scene) were recorded while autopilot was active. Eye glance reduction was manually coded. Average duration of a single glance away from road for all participants was 1.66 s, with a standard deviation of 3.60s. Typically, glances away from the road included glances to the cluster, center console, phone, or a food item. Average single glance time away from road per participant ranged from 1.20s to 3.20s. Average single glance away from road per participant while autopilot was active is shown in Figure 1.

To compare driver eye glance behavior while autopilot was engaged to known driver eye glance behavior while driving manually, a frequency distribution of all eyes off-road glance durations observed while the vehicle was in autopilot mode was plotted, as shown in Figure 2.

The frequency distribution plot indicated a mean duration of a single glance away from road was 1.66 s, while the median duration of a single glance away from road was 1.06 s. The 85th and 95th percentile duration of a single glance away from road were 2.41 s and 4.14 s respectively.

These results were comparable to single glance durations observed while tuning a car radio in manual driving as reported by [8] (see Figure 3). The mean off-road single glance duration in autopilot was somewhat longer than the mean single glance in a radio tuning task while driving manually (1.66 s vs. 1.44 s, respectively). However, the variability in single glance durations is roughly 7 times greater during autopilot driving compared to driving manually and tuning a radio (standard deviation of 3.6 s vs. 0.5 s, respectively).

Similarly, the results from eye off-road glance duration from the current study is consistent with results obtained by [9], which reported a median single off-road glance

FIGURE 1 Average duration of a single glance away from road while autopilot is active.

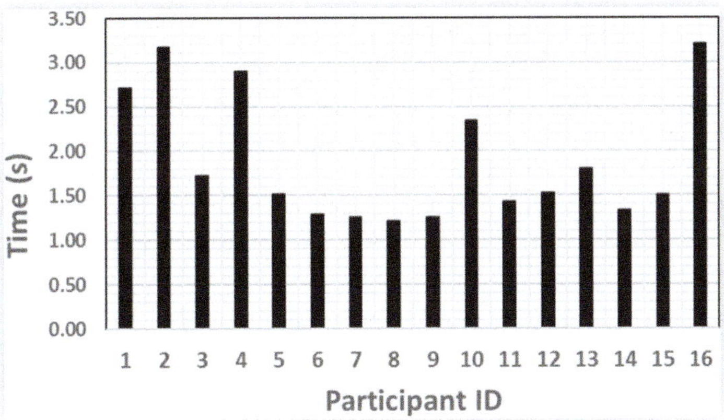

FIGURE 2 Driver eyes off-road glance distribution while autopilot is engaged.

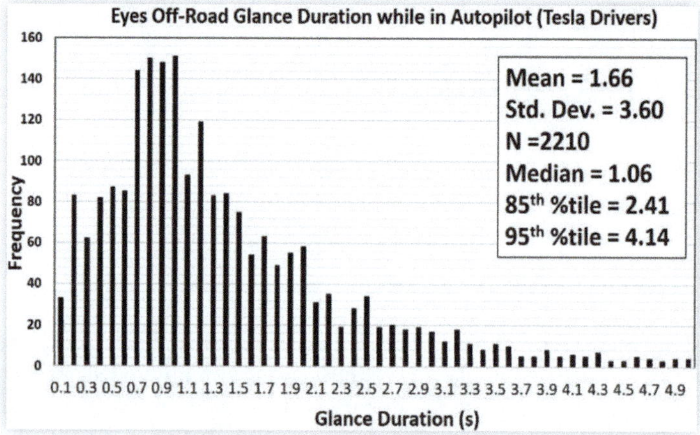

FIGURE 3 Distribution of eye glance duration when manually tuning a radio (Source Rockwell, 1988).

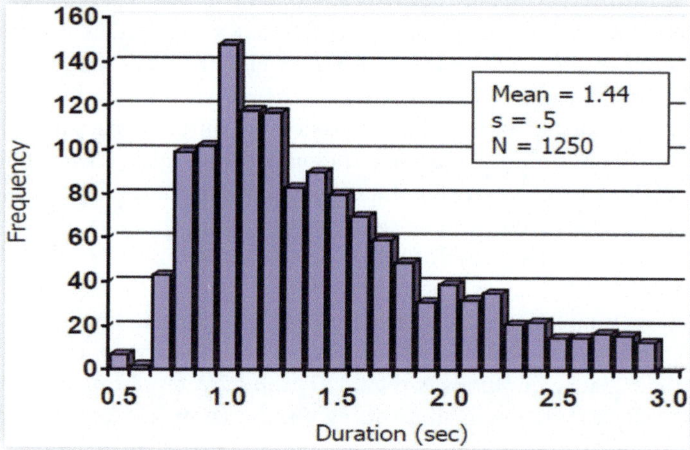

durations 2.18 s and 3.61 s respectively in a scheduled versus reference conditions in dynamic simulator environment.

Further, analysis of eye glance data showed that participants kept their eyes on the road about 89% of the total time that autopilot was active. Again, this finding from the current study is consistent with results reported by [10], which found that in a car following experiment, participants had their eyes on-road 86% of the time.

Thus, contrary to the perception that drivers may disengage when operating a partially automated vehicle, results from this study showed that drivers stayed alert and generally kept their eyes on the road. The results showed that typical (average) eyes off-road duration while autopilot was engaged is consistent with eyes off-road duration while tuning a radio in manual driving, though variation was substantial in autopilot driving.

Hands-On/Off Steering Wheel

Driver hands-on/off steering wheel while autopilot was active was compared to eyes off-road behavior and experience level of the driver. Table 1 below provides a summary

TABLE 1 Driver experience, eyes on-road, and hands-off steering wheel while autopilot is active

Participant ID	Length of Experience	% Eyes on-Road while in Autopilot	% Hands-Off while in Autopilot
5	3 weeks	93.1	1.6
15	3 weeks	93.9	7.3
9	less than 2 month	84.3	0
16	less than 2 month	78.8	40.4
7	2-4 months	79.2	72.6
11	2-4 months	93.3	99
4	4-8 months	86.5	9.5
6	4-8 months	96.5	4
8	4-8 months	96.7	0.7
3	8m - lyr	97.0	0.3
12	8m - lyr	97.1	51.6
14	8m - lyr	87.9	3.2
1	1 yr +	78.6	11.5
2	1 yr +	86.1	96.5
10	1 yr +	91.3	93.2
13	1 yr +	91.4	0.6
Average		**89**	**28**

of driver experience, eyes on-road and hands-off steering wheel behavior while autopilot was active.

The results showed that the level of autopilot usage experience did not influence percentage eyes on/off-road or percentage hands-on/off steering wheel. Participants' level of experience and percentage hands-off steering wheel is shown in Figure 4.

In addition, the results indicated that driver hands-on-wheel behavior is bi-modal. Eleven drivers kept at least one hand on the wheel less than 50% of the time, while the remaining five drivers kept one or both hands off the wheel 50% of the time or more while Autopilot was engaged. In essence, there were broadly two types of drivers: "*active drivers*" and "*supervisor drivers*". Regardless of the length of autopilot experience, *active drivers* kept their hands on the steering wheel for most of the time while autopilot was active. On the other hand, *supervisors* kept their hands-off the steering wheel most of

FIGURE 4 Participants' autopilot experience vs. percentage of hands-off while in autopilot.

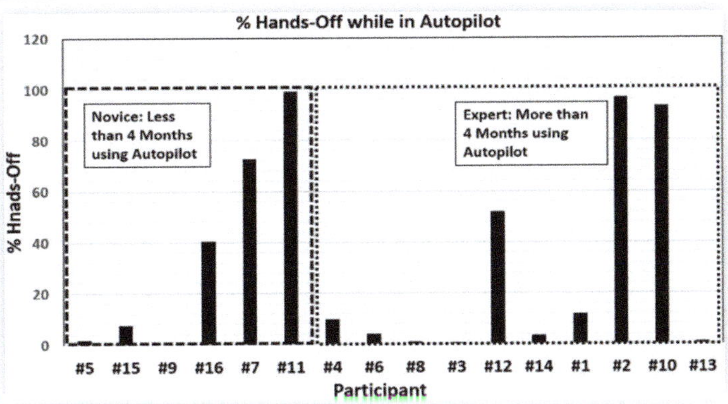

FIGURE 5 Total time in autopilot vs. total time on-highway and total hands-off time.

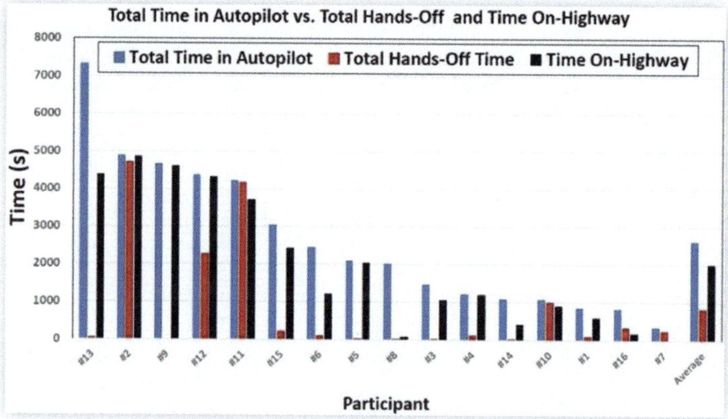

the time while autopilot was active. When asked, *active drivers* said they kept their hands on the steering wheel to: (i) ensure they were ready to take control when necessary, and (ii) avoid receiving hands-off warning from the system. On the contrary, *supervisors* said they kept their hands off the steering wheel because: (i) they felt comfortable and confident in the system, and (ii) they did not want to accidentally disengage autopilot. The authors believe this behavior is attributable to individual differences and acceptance of automation.

On-Highway/Off-Highway Autopilot Usage

Results also showed that participants were more likely to use autopilot while driving on highways (i.e. freeways). Specifically, participants used autopilot an average of about 70% on highway drives/trips. However, on non-highway trips (i.e. on secondary roads), participants used autopilot only about 7% of the time. This finding was not surprising as most participants had indicated earlier that they use autopilot in areas with visible lane lines/markings. This is one indication that the drivers developed some understanding of autopilot operation for lane keeping. Figure 5 shows total time in autopilot versus total hands-off time and time on-highway.

Common driver deactivation of autopilot while driving on a highway included lane changes, approaching slow vehicles, lane merging or lane splitting, and exiting highways. On secondary roads, drivers deactivated autopilot when they wanted to change lanes, make a turn, or when approaching stop sign, traffic light, or stopped/slow vehicles.

Tasks Performed While Autopilot Is Active

Overall, participants engaged in a variety of tasks while autopilot was active during the observation period. However, these tasks were no different from the tasks they engaged in while autopilot was off. Typical tasks performed while autopilot was active include: (i) sending and receiving text messages, (ii) making and receiving phone calls, (iii) interacting with center console (navigation, music etc.), (iv) interacting with steering wheel buttons, (v) eating and drinking, (vi) talking to passengers. When performing non-driving related tasks, participants periodically looked at the road, splitting their visual attention until the task was complete. This is evident in the eye glance data discussed above. Most participants reported that they were comfortable looking away from the road somewhat longer because they were confident autopilot would act as "safety buffer" while their eyes were off the road.

Summary/Conclusions

The results reported in this study lead to several observations. First, participants felt responsible for the operation of their vehicles when autopilot was active because they did not think autopilot had the capability to drive without supervision. Second, eye glance analyses from the study showed that participants' eye on/off road glance behavior while operating a partially automated vehicle (i.e. Tesla autopilot) was somewhat similar to eye off-road glance behavior in radio tuning task while driving manually though substantially more variable. Nonetheless, drivers while in autopilot mode glanced away from the road scene no longer than about 4 s in the vast majority of instances. This finding is significant, since it contradicts the common perception that drivers will disengage from monitoring when operating a partially automated vehicle. Third, analysis of eye glance data and hands on/off steering wheel data showed that eyes on/off road glance behavior had no relation to whether drivers kept their hands on the steering wheel or not. Further, the level of experience (how long a driver has used autopilot) did not influence eye off-road glance behavior, neither did it influence hands on/off steering wheel behavior. The results indicated a bi-modal driver behavior when it comes to the operation of partially automated vehicles as observed in this study: those who mostly kept their hands on steering wheel while autopilot was active (i.e. *active drivers*) and those who mostly kept their hands off the steering wheel while autopilot was active (i.e. *supervisors*).

Study Limitations

This study required drivers to manually activate/turn on the video/audio capture system before each trip. Therefore, drivers were generally aware that they were being recorded during these trips, this could have influenced them to exhibit more appropriate behavior than they normally would.

Second, the study used drivers who owned Tesla vehicles, and may not be a good representation of the general driving population.

References

1. Claypool, H., Bin-Nun, A., and Gerlach, J., "Self-Driving Cars: The Impact on People with Disabilities," The Ruderman Family Foundation, Ruderman White Paper, 2017.
2. SAE International On-Road Automated Vehicle Standards Committee, Standard J3016, "Taxonomy and Definitions for Terms Related to On-Road Motor Vehicle Automated Driving Systems," 2014.
3. National Highway Safety Transportation Administration, "Traffic Safety Facts," 2015.
4. Llaneras, R.E., Cannon, B.R., and Green, C.A., "Strategies to Assist Drivers in Remaining Attentive," 2016.
5. "While under Partially Automated Driving: Verification of Super Cruise Human Machine Interface Concepts."

6. Endsley, M.R. and Kiris, E.O., "The Out-of-the-Loop Performance Problem and Level of Control in Automation," *Human Factors* 37, no. 2 (1995):381-394.
7. Lerner, N., Singer, J., and Huey, R., "Driver Strategies for Engaging in Distracting Tasks Using In-Vehicle Technologies," Publication No HS DOT 810919, NHTSA, U.S. Department of Transportation, 2008.
8. Rockwell, T.H., Spare Visual Capacity in Driving-Revisited Gale A.G. et al. eds., *Vision in Vehicles II* (Amsterdam, Elsevier, 1988), 317-324.
9. Blommer, M., Curry, R., Kochhar, D., Swaminathan, R. et al., "Off-Road Glance Behavior, Response Time to a Forward Collision Hazard, and Engagement Strategy Effects in Automated Driving," *Proceedings of the Human Factors and Ergonomics Society Annual Meeting*, Austin, TX, 2017.
10. Tijerina, L., Barickman, F.S., and Mazzae, E.N., "Driver Eye Glance Behavior During Car Following," Publication No DOT HS 809 723, NHTSA, U.S. Department of Transportation, 2004.

CHAPTER 5

Forward Collision Warning: Clues to Optimal Timing of Advisory Warnings

Nazan Aksan
University of Iowa Hospital and Clinic

Lauren Sager, Sarah Hacker, Robert Marini, Jeffrey Dawson, and Steven Anderson
University of Iowa

Matthew Rizzo
University of Nebraska Medical Center

We examined the effectiveness of a heads-up Forward Collision Warning (FCW) system in 39 younger to middle aged drivers (25-50, mean = 35 years) and 37 older drivers (66-87, mean = 77 years). The warnings were implemented in a fixed based, immersive, 180 degree forward field of view simulator. The FCW included a visual advisory component consisting of a red horizontal bar which flashed in the center screen of the simulator that was triggered at time-to-collision (TTC) 4 seconds. The bar roughly overlapped the rear bumper of the lead vehicle, just below the driver's line-of-sight. A sustained auditory tone (~80 dB) was activated at TTC = 2 to alert the driver to an imminent collision. Hence, the warning system differed from the industry standard in significant ways. 95% Confidence intervals for the safety gains ranged from −.03 to .19 seconds in terms of average correction time across several activations. Older and younger adults did not differ in terms of safety gains. Closer inspection of data revealed that younger to middle aged drivers were already braking (42%) on a larger proportion of FCW activations than older drivers (26%), $p < .001$. Conversely, older drivers were still accelerating (38%) on a larger proportion of FCW activations than younger to middle aged drivers (23%) at the time FCW was activated, $p < .009$. There were no differences in the proportion of activations when drivers

CITATION: Aksan, N., Sager, L., Hacker, S., Marini, R. et al., "Forward Collision Warning: Clues to Optimal Timing of Advisory Warnings," *SAE Int. J. Trans. Safety* 4(1):2016, doi:10.4271/2016-01-1439.

were coasting at the time FCW was activated, p = .240. Furthermore, large individual differences in basic visual, motor, and cognitive function predicted the tendency to brake prior to FCW activation. Those who tended to be better functioning in each of these domains were more likely to be already braking prior to FCW activation at the fixed threshold of TTC = 4. These findings suggest optimal timing for advisory alerts for forward events may need to be larger than TTC = 4.

Introduction

Extending the mobility of a growing aging population safely has relevance to public health and injury prevention. Aging drivers are at increased risk for crashes [1] and emerging technologies that incorporate Advanced Driver Assist systems (ADAS) hold a particular promise in compensating for declining capacities of elderly to extend safe mobility in later years. Rear-end collisions are among the most common crashes in the US [2] and forward collision warning (FCW) systems are designed to reduce incidence of these types of crashes.

Relatively recent Insurance Institute for Highway Safety (IIHS) data suggest that vehicle models equipped with autonomous braking are associated with reduction in insurance claims. Data on similar vehicle models without autonomous braking feature suggest benefits of FCW maybe more limited [3]. One of the challenges in designing a FCW without the added benefits of autonomous braking is to determine the timing of the initial alert. Ideally, the timing of the alert would adjust for or factor in both the risk tolerance preferences and the capabilities of the driver in responding to a forward event. For example, aging drivers would be expected to react more slowly than younger to middle aged drivers owing to declines in visual, motor, and cognitive function typical of aging [4], hence alerts that would be effective for older drivers would have to be presented earlier.

Our goal in this study was to evaluate the effectiveness of an idealized FCW that differed from industry standards in significant ways especially with respect to its visual components, examine individual differences in profiles of pedal behavior at the time of FCW activation, and ask whether basic visual, motor, and cognitive functioning predicted pedal behavior and effectiveness as measured by reaction time to turn FCW off. Figure 1 illustrates the FCW we implemented in a fixed-based simulator.

FIGURE 1 Illustration of the FCW visuals. As the participant nears a vehicle ahead (depicted left to right), the size of the 'red' visual rectangle increases as a function of the increase in time-to-collision (TTC) value. When TTC $\leq$ 2.0 seconds, the visual is at its largest and is accompanied by a beeping tone to alert the driver to imminent danger. FCW system was only active at speeds above 5 mph to avoid over-alerting when approaching other vehicles at an intersection or in low speed traffic.

The *advisory* component of the FCW consisted of a red horizontal bar appearing in the center channel. The spatial position of the bar roughly overlapped the rear bumper of the lead vehicle, allowing the observer's attention to be rapidly drawn to the area of the environment where corrective action is needed. We elected to manipulate length and width of the visual icon because it creates a powerful sense of looming. This form of motion has been demonstrated to be a powerful attractor of attention in the literature [5]. A combination of driver manual recommendations and prior research [6, 7, 8] suggested a TTC of 4 seconds may be large enough to accommodate the declining reaction time of older drivers to forward-critical events. Hence the visual warning appeared when TTC reached 4 seconds. When TTC was reduced to 2 seconds or less, a sustained auditory tone was added to indicate an *imminent* crash.

To address the question of safety gains, younger to middle aged drivers and older drivers completed two versions of the same drive once with warnings turned on and once with warnings turned off. We included measures of both average reaction time (RT) to increase distance to greater than TTC = 4 and turn off FCW, total time drivers were exposed to FCW, and number of times they activated FCW. In addition, for each activation of FCW we tracked pedal behavior of the drivers as being in one of the three states: already braking, already costing, or still accelerating. These measures were tracked identically whether the participants actually saw the FCW or not. To gauge the appropriateness of TTC=4 threshold and functional significance of pedal behavior at the time of FCW activation, we examined these driving measures in relation to extensive assessments of basic visual, motor, and cognitive functioning.

Methods

Sample

The sample consisted of 39 younger to middle-aged and 37 older drivers. The younger to middle aged drivers were 25 to 50 years old (mean = 35 years), and consisted of 18 males and 21 females. The older drivers were 66 to 87 years old (mean = 77 years), and consisted of 23 males and 14 females. Inclusion criteria included: a) valid driver's license, b) minimum 10 years of driving experience, c) driving at least 1 hour or 50 miles per week, d) negative screen for dementia (MoCA > 18) [9], and e) corrected visual acuity better than 20/50. 35% of the sample scored in the Mild Cognitive Impairment range and the remaining scored normal on MoCA. 86% of the sample was white. Educational achievement was distributed as follows: 7% had high school or less, 45% had less than college degree, 45% had a college degree or more, others did not indicate education level.

Procedure and Design

Following basic visual function and dementia screening, participants took a simulator drive (fixed base, full cabin with 180° FOV, DriveSafety RS600) to assess motion sickness. Of those tested in the simulator, 28 got motion sick and were not part of the 76 whose data were analyzed here. The remainder participated in detailed assessments of their cognitive function including processing speed, memory, and visuospatial construction during a second visit (see Figure 2). In four additional visits scheduled 2 weeks apart, participants took eight simulated drives that took 15-18 minutes to complete designed to examine long-term adaptation to warning systems that included in addition to FCW, also a lane deviation warning and lane change/merge warning. The data from the other

FIGURE 2 Examples of tasks used to measure each of the cognitive function domains; processing speed, visuospatial construction, memory, and executive function.

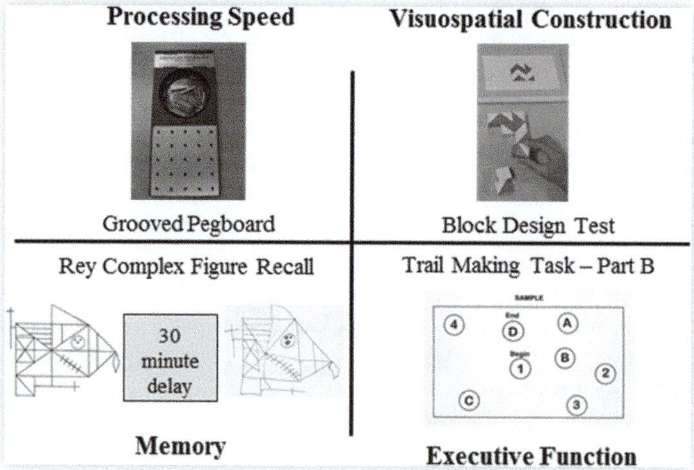

warning systems were not examined in this report. This report also did not address long-term adaptation to FCW. In each of the four visits, participants were exposed to the warnings-on version and warnings-off version of the same drive (i.e. same layout, with same hazards differing only in whether the warnings were turned off (not visible to driver) or on (visible to driver). The order of presentation was randomized and counterbalanced both with respect to: a) whether they experienced the no-warn or the warn version of the drive first in each of four visits b) which of the four distinct drive layouts they first experienced. The drive layouts were similar in terms of tile composition with respect to road culture, total time (15-18 minutes each), and number of hazards encountered.

Measures and Statistical Analyses

ASSESSMENTS OF FUNCTIONING

A battery of standardized tests measuring basic visual function [10, 11], motor function [12, 13, 14], and extensive assessment of cognitive functioning were administered [15, 16]. The assessments took approximately 2.5 to 3 hours. Figure 2 illustrates some of the tests in the cognitive battery.

Table 1 lists the specific tests included in the basic visual and motor function domains. Table 2 lists the specific tests that were included in each of the four cognitive functioning domains: processing speed, visuospatial construction, memory, and executive functioning. The assignment of the tasks to these four cognitive domains was guided by a confirmatory factor analytic study [17]. The raw scores obtained from the each of the tests were standardized with zscore transforms and then averaged to form composite scores following prior practices.

TABLE 1 A listing of the specific tests used to measure functioning in basic visual and motor function

	Tests
Visual Function	Near and far visual acuity, contrast sensitivity
Motor Function	Get-up & Go, Functional Reach

DRIVING PERFORMANCE MEASURES

The following measures were extracted from the drive files: Total exposure to FCW in a given drive, number of FCW activations, time it took participants to turn FCW off in seconds (correction RT). Number of activations, and time it took a participant to turn off FCW were computed in each drive and then averaged across all available drives. In addition, participants pedal

TABLE 2 A listing of the specific tests used to measure functioning in each of the domains of cognitive functioning

Domains of cognition	Tests
Processing Speed	Grooved Pegboard, Useful Field of View Total score, Trail Making Test-Part A
Visuospatial construction	Complex Figure-Copy, Line Orientation, Block Design
Memory	Controlled Oral Word Association, Complex Figure-Recall, Auditory Verbal Learning Test-30 minute Recall
Executive Function	Trail Making Test -Part B, Wisconsin Card Sort Test perseveration errors

behavior at the time FCW was activated was tracked as being in one three states: already braking, already coasting, or still accelerating. The pedal behavior was used to compute a proportion score reflecting the percentage of FCW activations in which the participant was in one of these three states to quantify individual differences in headway keeping. These proportion scores were also averaged across all available drives.

Results

A 2 × 2 mixed ANOVA was performed on each of these driving performance measures to examine in detail both the effectiveness of FCW and pedal use patterns to maintain headway for younger to middle aged drivers and older drivers. Warning status served as the within-subject factor, and age group as the between subject factor. The descriptive statistics from the cells of this design are shown in Table 3 through Table 8.

The 2X2 mixed design ANOVA on Total Exposure to FCW did not show any significant main effect of warning status, $p = .305$, or a main effect of age group, $p = .778$, or

TABLE 3 Descriptive statistics for each cell in the study design for total exposure measure

	Warnings On M (SD)	Warnings Off M (SD)	
Younger to middle aged drivers (n=39)	7.962 (3.26)	8.139 (3.62)	8.051 (3.04)
Older drivers (n=37)	7.589 (3.23)	8.117 (3.21)	7.853 (3.04)
	7.780 (3.23)	8.129 (3.41)	

TABLE 4 Descriptive statistics for each cell in the study design for number of FCW activations

	Warnings On M (SD)	Warnings Off M (SD)	
Younger to middle aged drivers (n=39)	5.581 (1.65)	5.730 (2.27)	5.657 (1.97)
Older drivers (n=37)	6.349 (2.61)	6.220 (2.03)	6.285 (1.97)
	5.950 (2.19)	5.97 (2.15)	

TABLE 5 Descriptive statistics for each cell in the study design for time to turn FCW off (correction RT)

	Warnings On	Warnings Off	
	M (SD)	M (SD)	
Younger to middle aged drivers	1.449	1.516	1.483
(n=39)	(0.37)	(0.46)	(0.32)
Older drivers	1.295	1.389	1.342
(n=37)	(0.39)	(0.37)	(0.32)
	1.374 (0.38)	1.454 (0.42)	

TABLE 6 Cell statistics for percent of activations when drivers were already braking at the time FCW was activated

	Warnings On	Warnings Off	
	M (SD)	M (SD)	
Younger to middle aged drivers	42.254	37.378	39.80
(n=39)	(17.16)	(16.72)	(13.11)
Older drivers	26.787	25.560	26.20
(n=37)	(14.89)	(17.79)	(13.38)
	34.724 (17.79)	31.625 (16.79)	

TABLE 7 Cell statistics for percent of activations when drivers were already coasting at the time FCW was activated

	Warnings On	Warnings Off	
	M (SD)	M (SD)	
Younger to middle aged drivers	33.582	35.680	34.60
(n=39)	(15.50)	(0.15)	(12.49)
Older drivers	30.517	31.987	31.30
(n=37)	(14.24)	(17.82)	(12.17)
	32.090 (14.88)	33.882 (16.64)	

TABLE 8 Cell statistics for percent of activations when drivers were still accelerating at the time FCW was activated

	Warnings On	Warnings Off	
	M (SD)	M (SD)	
Younger to middle aged drivers	21.422	24.900	23.20
(n=39)	(15.74)	(11.65)	(12.49)
Older drivers	39.036	37.650	38.30
(n=37)	(19.53)	(18.19)	(12.77)
	29.997 (19.68)	31.110 (16.39)	

an interaction effect of warning status by age group, $p = .611$. As can be seen from the cell means in Table 3, total exposure across 15-18 minute drives was very brief, 95% Confidence Intervals (CI) for the measure ranged from 7 to 9 seconds in both warnings-on and off drives. Older driver exposure was not different than the younger to middle aged drivers, and FCW affected both groups similarly on this measure.

FIGURE 3 Bar graphs for correction RT or time it took drivers to turn FCW off.

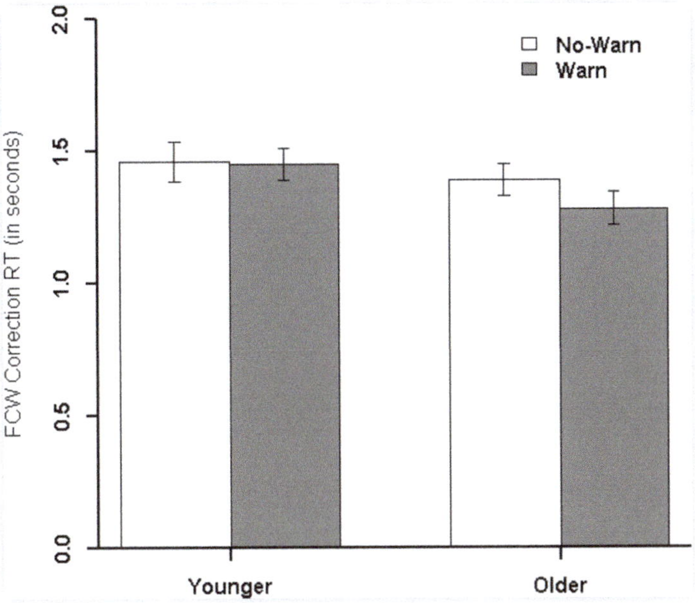

The 2X2 mixed design ANOVA on number of FCW activations did not show any significant main effect of warning status, p = .961, or a main effect of age group, p = .170, or an interaction effect of warning status by age group, p = .525. Table 4 shows cell statistics. 95% CIs indicated that FCW was activated between 5.5 and 6.5 times. Patterns of FCW activation was similar for both age groups whether warnings were turned on or not.

The 2X2 mixed design ANOVA on time it took the participants to turn FCW off did not show any significant main effect of warning status, p = .163, or an interaction effect of warning status by age group, p = .807. But there was a marginally significant effect of age group, p = .059. Table 5 shows cell statistics. 95% CIs indicated that it took between 1.23 to 1.55 seconds to turn FCW off whether participants saw the warnings or not. The marginally significant main effect showed an unexpected pattern in that older drivers appeared to turn off FCW more quickly than younger to middle aged drivers. Figure 2 presents bar graphs for those cell statistics.

The 2X2 ANOVAs on pedal behavior measures indicated significant differences in how the two age groups were maintaining headway at the time FCW was activated. Table 6 shows the cell statistics for the tendency to brake at the time FCW was activated and Figure 4 depicts them as a bar graph. There was a significant age difference in the tendency to brake at the time FCW was activated, p <.001. Younger to middle aged drivers were already braking at FCW activation 39.8% of the time (95% CI ranged from 35.6% to 44.0%) whereas older drivers were braking on 26.2% of the time at FCW activation (95% CI ranged from 21.9% to 30.5%). The tendency to brake at the time FCW was activated did not change as a function of whether the warnings were turned on and visible to drivers or not, p =.144. The age groups did not respond differently when FCW was turned on or off, p = .380.

The 2X2 ANOVA for the tendency to be coasting at the time FCW was activated did not show significant effects of age group, p = .240, warning-status p = .428, or an interaction effect between age group and warning status, p = .889. Figure 5 shows the bar graphs and corresponding cell statistics are shown in Table 7. Younger to middle aged drivers were already coasting at FCW activation 33.8% of the time (95% CI ranged from 30% to 37.6%) and older drivers were also already coasting 32% of the time (95% CI ranged from 28.6% to 35.5%).

FIGURE 4 Bar graphs for percent of activations when drivers were already braking at the time FCW was activated.

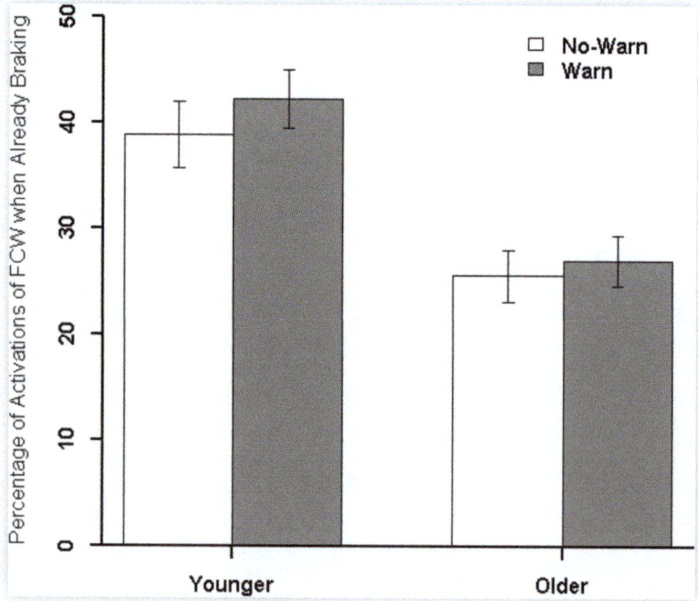

The 2X2 ANOVA for the tendency to be still accelerating at the time FCW was activated did not show significant effects of warning-status $p = .672$ or an interaction between age group and warning status, $p = .326$. However, there was significant main effect of age group, $p < .001$. Figure 6 shows the bar graphs and Table 8 shows the cell statistics from the design. There was a significant tendency for older adults to be still accelerating at the time FCW was activated compared to younger to middle aged drivers.

FIGURE 5 Bar graphs for percentage of activations when drivers were already coasting at the time FCW was activated.

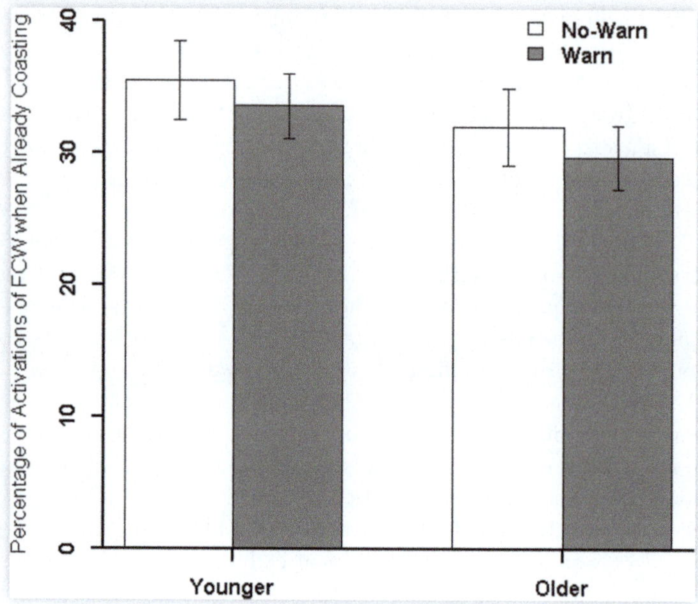

FIGURE 6 Bar graphs for percentage of activations when drivers were still accelerating the time FCW was activated.

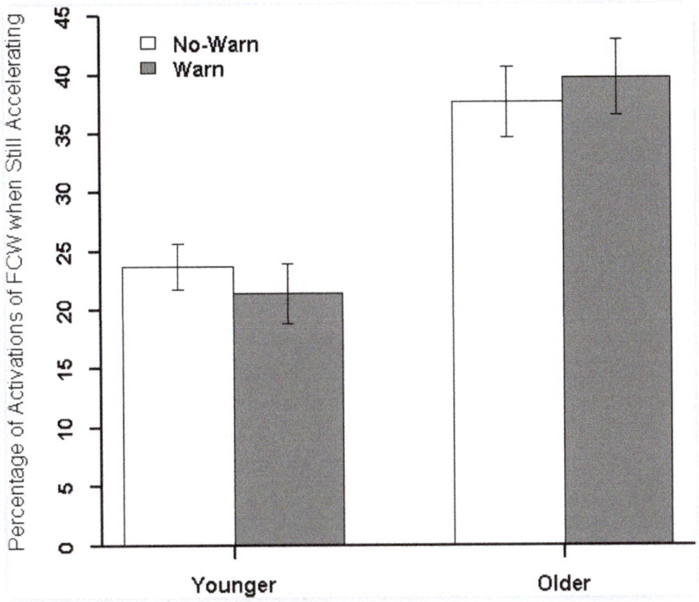

Older drivers were still accelerating 38.3% of the time (95% CI ranged 34.2% to 42.4%) at the time FCW was activated while younger to middle aged adults were found to be accelerating 23.2% of the time (95%CI ranged from 19.2% to 27.2%).

To further explain the meaning and significance of the individual differences observed in pedal behavior at the time of FCW activation we examined pedal behavior in relation to measures of basic visual, motor, and cognitive functioning. Table 9 presents the Pearson correlations for the whole sample in both the warnings-on and warnings-off version of the drives. As can be seen from Table 9, tendency to brake at the time FCW was activated was associated with better functioning while the tendency to be still accelerating at the time FCW was activated was associated with lower functioning in general. This pattern held whether the driving measures were obtained when the warnings were turned on and visible to participants or not.

TABLE 9 Pearson correlations among measures of functioning and pedal behavior at the time FCW was activated in both the warnings turned-on and off drives

Pearson r	Visual function	Motor function	Speed of processing	Memory	Visuospatial construction	Executive function
Warnings are off						
% time already braking	.309**	.303**	.373**	.235*	.180	.074
% time coasting	.031	.103	.132	.115	.079	.242*
% time still accelerating	-.267*	-.340**	-.452**	-.281*	-.224*	-.163
Warnings are on						
% time already braking	.401**	.257*	.469**	.357**	.154	.363**
% time coasting	.088	.103	.139	.218	.227*	.204
% time still accelerating	-.422**	-.360**	-.523**	-.497**	-.297**	-.433**

* p < .05.
** p< .01.

Summary/Conclusions

We evaluated the relative effectiveness of FCW for both older and younger to middle aged drivers that differed from industry standard in significant ways, especially with respect to its visual features. The *advisory* component of the FCW was visually distinct from objects in the driving scene overlapped with the rear bumper of the vehicle ahead, the length of the width of the visual icon was manipulated to create sense of looping to better capture attention between initial activation at TTC = 4 until the auditory chime was added at TTC =2 to signify an imminent crash.

The measures of FCW effectiveness including total exposure time, number of activations, or correction time did not indicate that FCW was effective in speeding either the deactivation of the warning, or reducing the total exposure. FCW also did not differentially affect older and younger to middle aged drivers. However, there was a trend suggested older drivers had slightly shorter correction times than younger to middle aged drivers. This was a marginally significant trend that was observed in both the warnings on and warnings off version of the drives. We believe this unexpected pattern was partly explained by the measures of pedal behavior around FCW activations. Older drivers tended to be still accelerating at the time FCW was activated and less likely to be braking compared to younger to middle aged drivers. Furthermore, the tendency to be still accelerating at TTC =4 was associated with lower functioning. This pattern of findings suggests that well-functioning drivers begin braking before the TTC =4 threshold was reached.

These findings suggest optimal timing for advisory alerts for forward events may need to be larger than TTC=4 especially for older drivers. We plan to conduct additional analyses to better identify the typical brake and coast times for older and younger to middle aged drivers to inform headway maintenance in general and typical TTC associated with those strategies to inform the timing of alerts. These additional analyses can also provide insight into the risk tolerance behavior of different age groups and whether risk tolerance varies as a function of cognitive functioning. Importantly, simulated drives while informative on relative rank-ordering of performance differences among age groups, may fail to generalize to the real-world in terms of absolute activation thresholds such as TTC= 4 is effective as simulated drives often lack the greater complexity of the real world.

Contact Information

Please address all correspondence to Nazan Aksan at
nazanaksan@uiowa.edu

Acknowledgments

1. The study was funded by Toyota-CSRC. We would like to thank James Foley and Kazutoshi Ebe for their guidance in designing both scenarios and input on LDW visuals.
2. The study would not be possible without he patience and perseverance of the participants in returning to the lab for repeated visits, and the hard work of the following research assistants: Amanda Farmer, Jessica Ferdig, Lacy Flanagan, Robert Marini, Tara Ohrt, and Sayeh Sabbagh.

Definitions/Abbreviations

SA - sample abbreviations
UBT - Use borderless table ≤ 3.5 inches wide.
test vector - Don't capitalize term unless an acronym or proper noun.

References

1. National Highway Traffic Safety Administration, "Traffic Safety Facts 2012: A Compilation of Motor Vehicle Crash Data from the Fatality Analysis Reporting System and the General Estimates System," DPT HS 812 032, 2012.
2. http://safebraking.com/nhtsa-28-of-crashes-are-rear-end-collisions/.
3. http://www.iihs.org/iihs/news/desktopnews/crash-avoidance-features-reduce-crashes-insurance-claims-study-shows-autonomous-braking-and-adaptive-headlights-yield-biggest-benefits.
4. Dawson, J.D., Uc, E.Y., Anderson, S.W., Johnson, A.M. et al., "Neuropsychological Predictors of Driving Errorsin Older Adults," *Journal of American Geriatric Society* 58 (2009):1090-1096.
5. Franconeri, S.L. and Simons, D.J., "Moving and Looming Stimuli Capture Attention," *Perception & Psychophysics* 65 (2003):999-1010.
6. Ayres, T.J., Li, L., Schleuning, D., and Young, D., "Preferred Time-Headway of Highway Drivers," *IEEE Intelligent Transportation Systems Conference Proceedings*, 2001.
7. Vogel, K., "A Comparison of Headway and Time to Collision as Safety Indicators," *Accident Analysis and Prevention* 35 (2003): 427-433.
8. Sultan, B. and McDonald, M., "Assessing the Safety Benefit of Automatic Collision Avoidance Systems (during Emergency Braking Situations)," *The 18th International Conference on the Enhanced Safety of Vehicles*, Nagoya, Japan, 2003.
9. Nasreddine, Z.S., Phillips, N.A., Bédirian, V., Charbonneau, S. et al., "The Montreal Cognitive Assessment (MoCA): A Brief Screening Tool for Mild Cognitive Impairment," *Journal of the American Geriatrics Society* 53 (2005): 695-699.
10. Pelli, D.G., Robson, J.G., and Wilkins, A.J., "The Design of a New Letter Chart for Measuring Contrast Sensitivity," *Clinical Vision Sciences* 2 (1988): 187-199.
11. Ferris, F.L. III, Kassoff, A., Bresnick, G.H., and Baily, I., "New Visual Acuity Charts for Clinical Research," *American Journal of Ophthamology* 94 (1982): 91-96.
12. Alexander, N.B., "Postural Control in Older Adults," *J Am Geriatr Soc* 42 (1994):93-108.
13. Duncan, P.W., Weiner, D.K., Chandler, J., and Studenski, S., "Functional Reach: A New Clinical Measure of Balance," *Journal of Gerontology* 45 (1990): M192-M197.
14. Mathias, S., Nayak, U.S.L., and Isaacs, B., "Balance in Elderly Patients: The "Get-up and Go Test"," *Archives of Physical Medicine and Rehabilitation* 67 (1986): 387-389.
15. Lezak, M.D., Howieson, D.B., and Loring, D.W., *Neuropsychological Assessment*, 4th ed. (New York: Oxford University Press, 2004).

16. Strauss, E., Sherman, E.M.S., Spreen, O.A., *Compendium of Neuropsychological Tests: Administration, Norms, and Commentary*, 3rd ed. (New York: Oxford University Press, 2006).

17. Anderson, S., Aksan, N, Dawson, J., Uc, E. et al., "Neuropsychological Assessment of Driving Safety Risk in Older Adults with and without Neurologic Disease," *Journal of Experimental and Clinical Neuropsychology* 34 (2012): 895-905.

CHAPTER 6

Situation Awareness, Scenarios, and Secondary Tasks: Measuring Driver Performance and Safety Margins in Highly Automated Vehicles

Madeleine Gibson, John Lee, Vindhya Venkatraman, Morgan Price, Jeffrey Lewis, Olivia Montgomery, and Bilge Mutlu
University of Wisconsin

Joshua Domeyer and James Foley
Toyota Technical Center USA, Inc.

The rapid increase in the sophistication of vehicle automation demands development of evaluation protocols tuned to understanding driver-automation interaction. Driving simulators provide a safe and cost-efficient tool for studying driver-automation interaction, and this paper outlines general considerations for simulator-based evaluation protocols. Several challenges confront automation evaluation, including the limited utility of standard measures of driver performance (e.g., standard deviation of lane position), and the need to quantify underlying mental processes associated with situation awareness and trust. Implicitly or explicitly vehicle automation encourages drivers to disengage from driving and engage in other activities. Thus secondary tasks play an important role in both creating representative situations for automation use and misuse, as well as providing embedded measures of driver engagement. Latent hazards-hazards that exist in the road environment and merit driver attention, but do not materialize to require a driver response-have been used with great success for understanding the vulnerability of novice drivers. Latent hazards might provide a similarly useful index of driver attention to the road during periods where the automation is vulnerable to failure. With highly automated vehicles, latent hazards include potential

CITATION: Gibson, M., Lee, J., Venkatraman, V., Price, M. et al., "Situation Awareness, Scenarios, and Secondary Tasks: Measuring Driver Performance and Safety Margins in Highly Automated Vehicles," *SAE Int. J. Passeng. Cars – Electron. Electr. Syst.* 9(1):2016, doi:10.4271/2016-01-0145.

roadway threats that might not be sensed by the automation and would require driver attention. This paper describes driving simulator scenarios used to operationalize automation-relevant latent hazards, secondary tasks tuned to index driver disengagement from the driving task, and measures that reflect safety margins rather than driving performance, such as drivers' trust, situation awareness, and expected time to transition to manual control.

Introduction

Sophisticated technology is already active in vehicle control. Driver assistance systems support lane keeping, parking, speed maintenance, blind spot monitoring, and they also enhance night vision and detect driver impairment. In general, the potential safety benefits of these safety systems and automated vehicles are promising; however, realizing this promise depends on carefully coordinating driver and vehicle behavior. As vehicles become more capable, this coordination can break down if the drivers' role is not clear.

Types and Levels of Automation

The types and levels taxonomy of automation [1, 2] describes how different sets of activities can be allocated to the automation and to what degree. For each stage of information processing from perception to control, levels of automation can range from none to complete. Taxonomies that guide design of vehicle automation [3, 4] share some similarities with the types and levels taxonomy: at one extreme the driver does everything and at the other the automation takes full control. The National Highway Traffic Safety Administration (NHTSA) and the Society of Automotive Engineers (SAE) define Level 1 vehicle automation as those that perform one primary vehicle control function - either steering or speed maintenance, while the driver performs the others, monitors the roadway, and remains fully responsible for safe vehicle control. Such automation is available in many vehicles today, for example, where the driver steers manually and engages cruise control. In NHTSA and SAE Level 2 automation, the automation performs at least two primary controls, for example both steering and speed maintenance. The driver is responsible for monitoring the roadway and should be ready to take-over control of the vehicle at any time [3]. Level 3 automation, from both NHTSA and SAE definitions, assigns higher capability to the automation, where the vehicle automation monitors the roadway and performs the primary control tasks; however, the driver may be expected to occasionally take control of the vehicle when the roadway demands exceed the capacity of the automation. The crucial concern for Levels 2 and 3 automation is the potential driver confusion regarding whether the driver or the automation is primarily responsible for driving. NHTSA Level 4 and SAE Level 4 and Level 5 represent automation that can handle all driving situations and requires very limited control input from the driver.

Both Level 2 and Level 3 automation require some amount of driver attention to the roadway, either continuously, at critical moments, or during transitions between levels of automation. Attention to the road for monitoring, compared to attention for control, leads to longer response times and less effective responses [5, 6, 7]. Perception for control involves visual, cognitive, and proprioceptive engagement that is coupled to drivers control inputs and the expected outcomes. Activating vehicle automation can sever the perception-action loop and transform the driver from a controller to a monitor. As a monitor, a driver is *on* or *out* of the control loop rather than *in* the control loop [8, 9]. Being *on* the control loop implies the driver is not perceiving the vehicle and roadway state to control the vehicle, but is actively monitoring the vehicle and roadway state to ensure the automation is controlling effectively. With automation that assumes control of steering and speed, drivers might easily disengage from monitoring and slip from on the loop to out of the loop [10].

Drivers might even think of driving as a distraction from other activities [11], and vehicle automation might be seen as a means of disengaging from driving, even when doing so violates the capabilities of the automation and compromises safety. Therefore, a major consequence of increasing automation might be the drivers' willingness to switch attention to non-driving tasks.

The evolving role of drivers in the context of automated driving and the additional activities drivers engage in present new challenges for evaluating how drivers work with the automation. Specifically, there are no standard scenarios and measures to evaluate joint control of the vehicle. This paper outlines a driving simulator evaluation protocol that captures driver engagement and disengagement in the driving task, and measures the safety margins achieved by drivers and automation.

Failure Modes of Driver-Automation Interaction

Central to assessing driver interaction with highly automated vehicles is the need to anticipate and test for likely failure modes. Risk analysis typically identifies failure modes associated with the mechanical, electronic, and software elements of systems and works to assure those risks remain below an acceptable level. Such risk estimates assume that drivers provide an additional safety margin, compensating for failures of the technology and other sources of unanticipated variability. This assumption is not always justified. Just as technology has failure modes, so do drivers and these failure modes should be assessed as part of a simulator-based evaluation.

Although highly automated vehicles lack operational exposure that might reveal prototypical driver-automation failure modes, experience in other domains suggests failure modes that might occur with vehicle automation. A very likely failure mode concerns drivers confusing Level 2 and Level 3 automation, where a critical distinction between these levels involves whether the driver has primary responsibility for monitoring the vehicle. *Responsibility diffusion* regarding whether the driver or the automation is primarily responsible for vehicle monitoring and control is likely a prominent failure mode. People are poorly suited to the role of monitoring automation and are prone to over trusting and neglecting reliable automation [12, 13]. *Mode confusion* associated with whether automatic control is engaged and if so what mode of control has been engaged is a prominent failure mode with automation in other domains and will likely affect vehicles equipped with combinations of Level 1, 2, and 3 automation [12, 14, 15, 16, 17]. *Operating envelope awareness*, similar to the more degraded situation awareness that can accompany automation introduction, concerns awareness of the intended operating envelope and proximity to safety boundaries of this operating envelope. This failure mode will affect vehicle automation that is not intended to be used on all road types and in all road conditions [18, 19, 20]. *Ineffective transfer of control* represents a general failure mode that might occur during planned, unplanned, and unwarranted transfers of control from the automation to the person [17, 21, 22]. Unwarranted transfers of control are those situations where the automation sees no need for driver intervention, but the driver initiates a steering or braking maneuver that might conflict with the automation. These automation failure modes define the requirements for assessing vehicle automation.

No single assessment protocol will likely address all the possible driver-automation failure modes. Several previous automated driving studies have focused primarily on the failure mode of ineffective transfer of control. The objective of these studies was to determine how quickly drivers recognize and respond to safety critical events. An alert or warning was used to prompt drivers to regain control of the vehicle when the automation fails. The method for how and when drivers were informed of needed actions was implemented differently across studies. In some cases the warning time was varied across conditions [23, 24]. Another study investigated driver response to one and two step processes to indicate a needed take-over [25]. This study also manipulated how drivers were alerted to the automation failure. The primary measure for these studies was the reaction time to begin the take-over process. Other driving performance metrics included minimum headway to lead vehicle, standard deviation of steering wheel position, and standard deviation of road offset (distance from the centerline of the road) for evaluating performance once the driver gained control.

Although many studies have focused on how and when control is traded, other automated vehicle studies have considered mode confusion failures and responsibility failure modes. These studies have examined driver behavior across different levels of automation. For example, manipulating the levels of automation (manual/fully automated) and traffic density (high/low) and drivers' willingness to overtake slow moving vehicles, management of car following, and secondary task engagement [26]. Similarly, drivers' response to automation was compared to an initial manual baseline condition [27]. When drivers experienced increasing levels of automation (i.e., lateral or longitudinal control followed by full automation) they were more willing to focus attention on secondary tasks as automation capability increased.

The following sections outline a protocol for addressing failure modes of responsibility diffusion, mode confusion, and operating envelope awareness. Another protocol is needed to address ineffective transfer of control. We describe latent hazards as an important element of driving simulator scenarios, secondary tasks as a central component of driving highly automated vehicles, and measures of safety margin, rather than driver performance, as a critical indicators of resilience in the face of driver-automation failure modes.

Driving Simulator Protocol

Driving simulators are usually composed of the following elements: cabs, computers and electronics, vehicle dynamics, scenario, and task environment [28]. Driving simulators have increasingly become a widely used and accepted tool for transportation human factors research due to the several advantages. One major advantage is the safe environment driving simulators provide to understand basic human limitations and driver behavior in safety critical events. For example, in distracted driving studies, experimenters can evaluate invehicle systems through driver engagement in secondary tasks in situations that would be dangerous on the road. Similarly, design of vehicle automation and in-vehicle technologies can be evaluated without the risk inherent in on-road and test track evaluations. Another advantage of driving simulators is the controlled environment. Each participant is exposed to identical driving scenarios, eliminating confounding variables such as weather or traffic found in naturalistic driving environments. Roadway conditions and other vehicle behaviors can be specified for the duration of the experiment. Lastly, driving simulators allow for drivers to experience many test conditions in a short time [29]. During a single study session, drivers can experience many road situations that might take hours or months to occur in naturalistic driving.

Although choosing the correct methods for conducting research is important, it is equally, important is choosing driving performance metrics that are sensitive to automation failure modes. Measures used to assess driver distraction with manual driving have included speed, vehicle following (headway), lane keeping, steering wheel metrics, event detection, response times, and subjective ratings. However, these measures do not consider joint performance of the driver and automation, such as when the driver is no longer in control of the vehicle's primary functions. With automated driving, the driver's role changes from being directly engaged in control to that of a monitor.

Drivers with automation need to be considered part of a joint cognitive system, with the unit of analysis moving from that of the driver to the driver-vehicle combination. Therefore, measures of driver performance should not focus simply on the driver, but on the joint performance of the automation and the driver. Furthermore, because vehicle automation can achieve very high levels of driving performance (e.g., maintain a fixed speed precisely) safety margins are more relevant. A major challenge is to measure safety margins that are maintained by the driver across different levels of automation and in response to a range of roadway situations. More specifically, this involves measuring driver awareness of the automation capability and driver adaptive capacity relative to automation limits.

Situation awareness and trust in automation are important indicators of drivers' adaptive capacity. Situation awareness is defined as "the perception of the elements in the environment within a volume of time and space, the comprehension of their meaning, and the projection of their status in the future" [30]. The first component of situation awareness (SA) has been measured as attention allocation [31], and can be estimated by drivers' glance behavior to road situations that might challenge the automation. Trust is an attitude that the automation will achieve the drivers' goals [32], and could be indicated by the degree to which drivers engage in non-driving tasks, neglect the roadway, and keep hands and feet away from the controls.

Latent Hazards to Assess Situation Awareness

Latent hazards offer a promising measure of situation awareness of automation. Latent hazards are possible roadway threats that do not necessarily develop into hazards that require immediate action [33]. Latent hazards have been used for understanding the limits of novice drivers. Novice drivers' attention to potential threats differs from that of experienced drivers [34, 35]. Attention to latent hazards is often measured using glance metrics. Drivers glances can be used as an indicator of what the driver is attention and what information the driver is processing. Whether drivers gaze towards latent hazards can indicate whether they will anticipate the potential threat and are ready to act if needed [36]. Latent hazards might provide a similarly useful measure of driver attention to the road situations where automation is less capable and when the driver needs to resume control.

A critical consideration in using latent hazards to evaluate situation awareness of vehicle automation concerns drivers being part of a joint cognitive system consisting of the driver and the automation. Previous use of latent hazards has focused on challenges that drivers must accommodate to maintain safety. When automation acts with drivers to control the vehicle, latent hazards need to be defined in terms of situations that challenge the automation and might require the driver to intervene.

We have designed scenarios to reflect situations that challenge automation. These scenarios were implemented as latent hazards. Such latent hazards involve potential

FIGURE 1 Scenario layout showing a possible ordering of latent hazards over a drive.

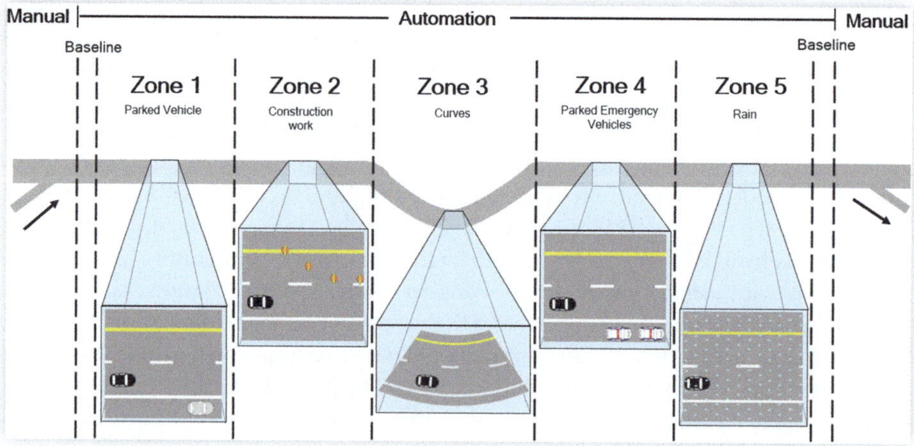

safety conflict situations that do not develop into active threats. However, drivers need to pay attention to the possibility of such hazards. Ideally, drivers are expected to recognize when latent hazards are present and be prepared to intervene to ensure safe operation of the vehicle. The hazards in our scenario do not require any control actions from the drivers. However, drivers can intervene at any point.

Figure 1 shows a drive composed of several scenarios. At the start of the drive, drivers manually operate the vehicle to merge onto the highway. When instructed, the drivers engage the automation, followed by one minute of baseline driving with the automation. A one-minute baseline period also occurs at the end of the scenario before switching back to manual control and exiting the highway. The rest of the drive includes five different zones. The zones are equal in distance and each includes one latent hazard. Latent hazards include: stopped vehicle on the side of the road, construction work in the adjacent lane, curves, emergency vehicles on the side of the road, and rain. The duration of the latent hazards is small relative to the period where the vehicle can easily accommodate roadway demands. For example, the construction work is 30 seconds in a zone of several minutes.

Measures for assessing drivers' response to the latent hazards focus on glance behavior: total glance duration to latent hazard, frequency of glance to hazard, and time of first glance to hazard [33, 36]. Beyond these measures, trust in automation might be reflected in behaviors that include moving hands to steering wheel, control inputs to the steering wheel, and foot movement towards the brake. The specific latent hazards considered in an evaluation depend on the design features of the automation. Latent hazards should represent situations at the edges of the operating envelope where relying on the automation is not appropriate or situations where drivers are expected to intervene in the event of an automation failure.

Secondary Task Engagement

A major motivation for developing vehicle automation and a major motivation for drivers engaging vehicle automation is the freedom such automation affords in performing secondary tasks. Such secondary tasks include email, social network interactions, and audio and video entertainment. Given the current engagement in secondary tasks while driving, automation evaluation should consider drivers interacting with a relatively engaging secondary task. In addition, secondary task engagement reflects drivers' trust in the automation and their willingness to neglect the monitoring of the road and automation. Attention to latent hazards measure the engagement in the driving task, secondary task usage measures disengagement from the driving task.

When selecting a secondary task to use in an automated driving study, it is important to choose one representative of the experience drivers will enjoy when using automated driving, such as a self-paced interaction with an information system. Many instances of self-paced tasks have been used in previous automated driving studies. These tasks include interaction with the in-vehicle entertainment system, eating, reading magazines, playing hand-held games, watching movies or TV shows, listening to the radio, performing grooming tasks, or completing word puzzles [24, 26, 27].

The secondary task protocol we developed consists of sorting emails into three categories, 1) Work, 2) Friends and Family, or 3) Trash, as described in Table 1. In the task, the driver sees a series of email subject lines displayed in a touchscreen application that models some basic functions of a common email client on a mobile device. Subject lines from the categories are chosen at random in a distribution of 1:1:4 respectively.

To complete the e-mail task, drivers need to press the edit button in the top right corner. This button prompts checkboxes to appear next to the subject lines. Drivers can

TABLE 1 Categories and examples of email messages

Email category	Example
Work	Emails relate to work, mentioning deliverables, and distinctly work related meetings or events. • Senior Developer Project Updates • Expense Reports Deadline Tuesday • August Departmental Schedule Request
Family and Friends	Emails relate to family affairs and social gatherings. • Family Dinner on Sunday at 4 • Grandma Birthday Celebration • Get Together 3/21 Downtown?
Trash	Emails are spam, featuring promotions, deals, and products. • Google TV, iPad, MacBook Air • Today Only: Save 20% on Holiday Gifts • Celebrate Summer with a Weekend Cruise

FIGURE 2 The main display shows the four emails available for sorting, a header with counts of the number of emails correctly sorted into each category, and the edit button.

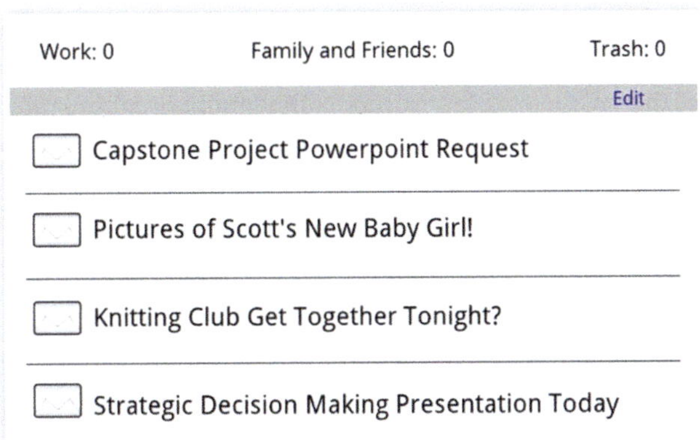

then select one or more e-mails to sort into the appropriate category. If the driver correctly sorts the email, the count on the top of the homepage is updated to reflect the total number of emails sorted for each category. If the email is incorrectly sorted, the count remains the same. On the sorting page, a cancel button returns that task to the homepage.

Because each subtask is comprised of several discrete actions that require input from the driver, the secondary task can serve as a surrogate for eye glance and driver engagement in the secondary task. To further ensure driver attention is directed to the task, a pop-up message appears if the task is inactive for more than five seconds. The message directs drivers to touch the screen to resume. Data are collected for each button press, and so the task precisely indexes driver engagement. Overall, this task provides an activity representative of what drivers might do in a highly automated vehicle and one that precisely records drivers' task engagement.

Measures of Driver Performance and Safety Margins

Driving simulator studies often measure driver performance in responding to events or maintaining vehicle control. These measures become less relevant in understanding driver interaction with highly automated vehicles. With measures of driving performance, the underlying assumption is that better driving performance corresponds to improved safety, which may not be the case with highly automated vehicles. Automation might maintain perfect control performance before it fails catastrophically. Because continuous measures of vehicle control, such as standard deviation of lane position, will likely fail to indicate the safety achieved by the joint cognitive system of the driver and automation, alternatives are needed.

With highly automated systems the distinction between performance and resilience is critical. Resilience represents the actions, time, and resources that enable a system to accommodate unexpected demands [37]. Automation often performs well, but is brittle, failing in the face of unexpected demands. Humans enhance the resilience of the system through their ability to adapt to the unexpected. A crucial measure is the degree of resilience the automation affords. This resilience can be measured in terms of safety margins. Safety margins reflect the capacity of the automation to respond to road situations and the response time of drivers to compensate for automation limitations. This response time can be estimated as a function of drivers' trust in the automation and situation awareness. Quantifying such safety margins represents an important measurement challenge in assessing vehicle automation.

This paper focused on simulator-based methods for evaluating vehicle automation. Equally important are analytic methods to evaluate automation and assess potential failure modes. Several promising techniques have emerged to support formal evaluation of human-automation interaction [38, 39, 40]. These complement simulator-based methods because the can uncover failure modes that might occur too rarely to be detected in a simulator evaluation and yet these failure modes might substantially undermine vehicle safety.

Conclusion

Highly automated vehicles will dramatically change the role of the driver. To ensure such changes enhance rather than degrade driving safety, an evaluation of design assumptions and driver-automation failure modes is needed. Driving simulators offer a promising approach to addressing these failure modes, but only if the scenarios, secondary tasks, and measures are tuned to the particular demands of vehicle automation assessment. Drivers and automation should be considered as part of a joint cognitive system that is vulnerable to new failure modes, and measures of safety margin rather than driving performance are most appropriate to ensuring these failure modes do not compromise driving safety.

Contact Information

John D. Lee can be reached at
john.d.lee@wisc.edu

Acknowledgments

Funding for this work was provided by Toyota Collaborative Safety Research Center (CSRC).

References

1. Parasuraman, R., Sheridan, T.B., and Wickens, C.D., "A Model for Types and Levels of Human Interaction with Automation," *IEEE Trans. Syst. Man Cybern. - Part A Syst. Humans* 30, no. 3 (2000): 286-297.
2. Sheridan, T.B. and Verplank, W., *Human and Computer Control of Undersea Teleoperations* (1978).
3. NHTSA, "Preliminary Statement of Policy Concerning Automated Vehicles," 2013.
4. SAE International Surface Vehicle Information Report, "Taxonomy and Definitions for Terms Related to On-Road Motor Vehicle Automated Driving Systems (J3016)," SAE Standard J3016, Rev. January 2014.
5. Eprath, A.R. and Young, L.R., Monitoring vs. Man-in-the Loop Detection of Aircraft Control Failures, Rasmussen, J. and Rouse, W.B. eds., *Human Detection and Diagnosis of System Failures* (New York: Plenum Press, 1981), 143-154.
6. Gibson, J.J., "Observations on Active Touch," *Psychol. Rev.* 69 (1962): 477-491.
7. Walker, G.H., Stanton, N., and Young, M.S., "The Ironies of Vehicle Feedback in Car Design," *Ergonomics* 49, no. 2 (2006): 161-179.
8. Bainbridge, L., "Ironies of Automation," *Automatica* 19, no. 6 (1983): 775-779, doi:10.1016/0005-1098(83)90046-8.
9. Wickens, C.D. and Kessel, C., "The Effects of Participatory Mode and Task Workload on the Detection of Dynamic System Failures," *IEEE Trans. Syst. Man. Cybern.* 9, no. 1 (1979):24-34, doi:10.1109/TSMC.1979.4310070.
10. Merat, N. and Lee, J.D., "Preface to the Special Section on Human Factors and Automation in Vehicles: Designing Highly Automated Vehicles With the Driver in Mind," *Hum. Factors J. Hum. Factors Ergon. Soc.* 54, no. 5 (2012): 681-686.
11. Hancock, P.A., Driven to Distraction and Back Again, Regan, M.A., Lee, J.D., and Victor, T.W. eds., *DRiver Distraction and Inattention: Advances in Research and Countermeasures*, 1st ed. (Surrey, England: Ashgate Publication Limited, 2013), 440

12. Bainbridge, L., "Ironies of Automation," *Automatica* 19, no. 6 (1983): 775-779, doi:10.1016/0005-1098(83)90046-8.
13. Strand, N., Nilsson, J., Karlsson, I.C.M., and Nilsson, L., "Semi-Automated Versus Highly Automated Driving in Critical Situations Caused by Automation Failures," *Transp. Res. Part F Traffic Psychol. Behav.* 27 (2014): 218-228, doi:10.1016/j.trf.2014.04.005.
14. Sarter, N.B., Woods, D.D., and Billings, C.E., Automation Surprises, Salvendy, G. ed., *Handbook of Human Factors and Ergonomics*, 2nd ed. (New York: Wiley, 1997), 1926-1943.
15. Sarter, N.B. and Woods, D.D., "How in the World Did We Ever Get into That Mode? Mode Error and Awareness in Supervisory Control," *Hum. Factors* 37, no. 1 (1995): 5-19.
16. Degani, A., Shafto, M., and Kirlik, A., "Modes in Automated Cockpits: Problems, Data Analysis, and a Modeling Framework," *Proceedings of the 36th Israel Annual Conference on Aerospace Science*, Haifa, Israel, 1995.
17. Merat, N. and Lee, J.D., "Preface to the Special Section on Human Factors and Automation in Vehicles: Designing Highly automated Vehicles with the Driver in Mind," *Hum. Factors J. Hum. Factors Ergon. Soc.* 54, no. 5 (2012): 681-686, doi:10.1177/0018720812461374.
18. Pritchett, A.R., "Aviation Automation: General Perspectives and Specific Guidance for the Design of Modes and Alerts," *Rev. Hum. Factors Ergon.* 5, no. 1 (2009): 82-113, doi:10.1518/155723409X448026.
19. Young, M.S., Stanton, N.A., and Harris, D., "Driving Automation: Learning from Aviation about Design Philosophies," *Internaltional J. Veh. Des.* 45, no. 3 (2007): 323-338.
20. Rasmussen, J., "Risk Management in a Dynamic Society: A Modeling Problem," *Saf. Sci.* 27 (1997): 183-213, doi:10.1016/S0925-7535(97)00052-0.
21. Merat, N., Jamson, A.H., Lai, F.C.H., Daly, M. et al., "Transition to Manual: Driver Behaviour when Resuming Control from a Highly Automated Vehicle," *Transp. Res. Part F Traffic Psychol. Behav.* 26, no. PART A (2014): 1-9, doi:10.1016/j.trf.2014.05.006.
22. Parasuraman, R. and Sheridan, T.B., "A Model for Types and Levels of Human Interaction with Automation," *Syst. Man* 30, no. 3 (2000): 286-297, doi:10.1109/3468.844354.
23. Gold, C., Dambock, D., Lorenz, L., and Bengler, K., "'Take Over!' How Long Does It Take to Get the Driver Back into the Loop?" *Proc. Hum. Factors Ergon. Soc. Annu. Meet.* 57, no. 1 (2013): 1938-1942, doi:10.1177/1541931213571433.
24. Mok, B., Johns, M., Lee, K.J., Miller, D. et al., "Emergency, Automation Off?: Unstructured Transition Timing for Distracted Drivers of Automated Vehicles," *Proc. 2015 IEEE 18th Int. Conf. Intell. Transp. Syst.*, 2015.
25. Miller, D., Sun, A., Johns, M., Ive, H. et al., "Distraction Becomes Engagement in Automated Driving," *Proc. Hum. Factors Ergon. Soc. Annu. Meet.* (2015): 2-6.
26. Jamson, A.H., Merat, N., Carsten, O.M.J., and Lai, F.C.H., "Behavioural Changes in Drivers Experiencing Highly-Automated Vehicle Control in Varying Traffic Conditions," *Transp. Res. Part C Emerg. Technol.* 30 (2013): 116- 125, doi:10.1016/j.trc.2013.02.008.
27. Carsten, O., Lai, F.C.H., Barnard, Y., Jamson, H. et al., "Control Task Substitution in Semi-Automated Driving: Does It Matter What Aspects Are Automated?," *Hum. Factors* 54, no. 5 (2012): 747-761.

28. Allen, R.W., Rosenthal, T.J., and Cook, M., A Short History of Driving Simulation, *Handbook of Driving Simulation for Engineering, Medicine, and Psychology* (2011): 2-3.

29. Young, K.L., Regan, M.A., and Lee, J.D., Measuring the Effects of Driver Distraction: Direct Driving Performance Methods and Measures, *Driver Dirstraction: Theory, Effects, and Mitigation* (2009): 86-102.

30. Endsley, M.R., "Toward a Theory of Situation Awareness in Dynamic Systems," *Hum. Factors J. Hum. Factors Ergon. Soc.* 37, no. 1 (1995): 32-64, doi:10.1518/001872095779049543.

31. Gugerty, L., Situation Awareness in Driving, *Handbook of Driving Simulation for Engineering, Medicine, and Psychology* (2011), 1-8.

32. Lee, J.D. and See, K.A., Trust in Computer Technology and the Implications for Design and Evaluation, Miller, C. Ed., *Etiquette for Human-Computer Work: Technical Report FS-02-02* (Menlo Park, CA: American Association for Artificial Intelligence, 2002), 20-25.

33. Vlakveld, W., Romoser, M.R.E., Mehranian, H., Diete, F. et al., "Do Crashes and Near Crashes in Simulator-Based Training Enhance Novice Drivers' Visual Search for Latent Hazards?," *Transp. Res. Rec. J. Transp. Res. Board* (2011): 153-160, doi:10.1016/j.biotechadv.2011.08.021.Secreted.

34. Garay-Vega, L., Fisher, D.L., and Pollatsek, A., "Hazard Anticipation of Novice and Experienced Drivers: Empirical Evaluation on a Driving Simulator in Daytime and Nighttime Conditions," *Transp. Res. Rec.* 2009, no. 1 (2007): 1-7, doi:10.3141/2009-01.

35. Underwood, G., Crundall, D., and Chapman, P., "Driving Simulator Validation with Hazard Perception," *Transp. Res. Part F Traffic Psychol. Behav.* 14, no. 6 (2011): 435-446, doi:10.1016/j.trf.2011.04.008.

36. Taylor, T., Pradhan, A.K., Divekar, G., Romoser, M. et al., "The View from the Road: The Contribution of on-Road Glance-Monitoring Technologies to Understanding Driver Behavior," *Accid. Anal. Prev.* 58, no. null (2013): 175-186, doi:10.1016/j.aap.2013.02.008.

37. Woods, D.D. and Branlat, M., "Hollnagel's Test: Being 'In Control' of Highly Interdependent Multi-Layered Networked Systems," *Cogn. Technol. Work* 12, no. 2 (2010): 95-101, doi:10.1007/s10111-010-0144-5.

38. Degani, A. and Heymann, M., "Formal Verification of Human-Automation Interaction," *Hum. Factors* 44, no. 1 (2002): 28-43.

39. Bolton, M.L., Bass, E.J., Member, S., and Siminiceanu, R.I., "Using Formal Verification to Evaluate Human-automation Interaction: A Review," 43, no. 3 (2013): 488-503.

40. Leveson, N.G., "A Systems-Theoretic Approach to Safety in Software - Intensive Systems," *IEEE Trans. Dependable Secur. Comput.* 1, no. 1 (2004): 66-86, doi:10.1109/TDSC.2004.1.

Frontal Collisions—What Are the Limitations of Future Forward-Looking Safety Systems?

Frank Keck, Marco Alt, Arne Vater, and Joseph Wessner
IEE Sensing Germany

Current driver assistance systems or forward-looking safety systems mainly address traffic scenarios with cars travelling in the same direction or being stopped. These scenarios are - considered from a dynamical point of view - comparatively easy to handle due to the limitations of the relevant scenario parameters (relative velocity, possible accelerations, ...). In the future it will be necessary to address oncoming traffic scenarios as well. These oncoming scenarios are responsible for a lot of critical accidents and the potential benefit is very high if one is able to reduce the crash severity in these scenarios. The problem remains that these scenarios are highly dynamical and therefore difficult to evaluate and handle.

The following questions are of interest:

- How should a system be designed to be able to handle these situations?
- What are the critical scenarios which define the performance of the whole system?
- What are the limitations which cannot be overcome?

We show a method based on simulation and data mining techniques which is able to address these questions. A stochastic approach based

CITATION: Keck, F., Alt, M., Vater, A., and Wessner, J., "Frontal Collisions—What Are the Limitations of Future Forward-Looking Safety Systems?," SAE Technical Paper 2017-26-0013, 2017, doi:10.4271/2017-26-0013.

on different use cases is used to generate the relevant data. The method described is generic and can also be used to evaluate, develop or improve other safety or driver assistance systems.

Introduction

In the last couple of years the first generations of advanced driver assistance systems or integral safety systems have entered the market and contributed positively to the safety of road usage. Nonetheless, it is a long way to go to reach Vision Zero, i. e. the avoidance of every death in traffic [1, 2]. We want to discuss which contributions current and future advanced driver assistance systems or integral safety systems can provide and where the limitations of these systems lie.

We will focus our discussion on forward collision avoidance systems, especially in head-on collision scenarios and will consider where the limits of these systems are reached. Despite the proven benefit of these systems [3], they still lack the ability to address every possible accident scenario. This is due to many factors: limited sensor information of the environment due to sensor shortages or latencies, unpredictable driver behavior, the stochastic nature of the evolution of a given traffic scenario, the high dynamics when oncoming traffic is considered, etc. We will address these uncertainties using a stochastic approach. As an introduction to these topics, we refer to the doctoral research studies [4, 5].

Similar approaches have been used by [6] to [11]. In [6], crash probabilities of planned paths are considered by computing stochastic reachable sets using Markov chains. The computation of collision risk levels based on Gaussian processes and Markov models to describe the evolution of systems can be found in [11]. A pseudo continuous evolution of probabilities of collisions is discussed in [10]. To improve computational efficiency, methods like importance sampling have been considered [9]. A completely different approach for the evaluation of driving prediction problems using dynamic Baysian networks can be found in [8].

The paper is structured in the following way: In the first section we describe a simple one dimensional system which can be solved analytically to show the intrinsic stochastic nature of the systems discussed. The basic notations like time to collision and time of unavoidability are introduced. Early considerations concerning time to collision can be found in [12]. The next section shows the transfer of these concepts to the two dimensional case. The analytical discussion is supported and illustrated using numerical simulations. To reduce the computational effort, we introduce the concept of limiting trajectories. After that, we present different scenarios in which the inevitable limitations of current forward collision avoidance systems are demonstrated. The paper finishes with considerations how these limitations might be overcome.

Limitations of Current and Future Pre-crash Systems

When developing a pre-crash function the following questions have to be addressed:

- What is the probability of a collision?
- When will it happen?
- Is there a possibility that we can avoid the collision?

Depending on the answer to these questions the algorithms decide what to do. We show how these questions can be answered for our model systems. The results can be adapted to real cars.

One Dimensional Example

We start by discussing a simple one dimensional system to introduce the basic ideas. We consider two cars moving on a one dimensional road. Because we do not have to take into account the lateral dynamics, the whole system can be described using relative coordinates (x_{rel}, v_{rel}, a_{rel}), where the different parameters denote relative distance, relative velocity and relative acceleration.

Furthermore, using a no change assumption during a scenario for the acceleration ($a_{rel} = const$), we can label every possible scenario depending on the initial values as part of the following set:

$$S = \{(x_{rel}, v_{rel}, a_{rel}) | x_{rel} \in \mathbb{R}, v_{rel} \in \mathbb{V}, a_{rel} \in \mathbb{A}_{v_{rel}}\} \tag{1}$$

where $\mathbb{V}$ and $\mathbb{A}_{v_{rel}}$ are the sets of possible relative velocity and acceleration values. Note that the set $\mathbb{A}_{v_{rel}}$ depends on the current relative velocity v_{rel}.

The whole time evolution is then defined by these initial parameters. For each scenario $s \in S$ we now consider whether there will be a crash in positive time ($t \geq 0$). We define the indicator or Heaviside function

$$y_s(t) = \Theta(t - t_c), t \geq 0 \tag{2}$$

which denotes for each scenario $s \in S$ if a crash has already happened ($y_s(t) = 1$, $t \geq t_c$) or not ($y_s(t) = 0$, $t < t_c$). The time t_c is called the time of collision. If it exists, it is unique for each scenario and given by a function T_c of the initial parameters:

$$t_c = T_c(x_{rel}, v_{rel}, a_{rel}). \tag{3}$$

In reality t_c is not a single value but can take on different values due to the fact that a given situation can evolve into different scenarios. The time of collision t_c then takes the value depending on a probability distribution $p_t(t_c)$.

To model this fact we consider different scenarios where the initial values of x_{rel} and v_{rel} are fixed and a_{rel} is sampled according to a predefined probability distribution over the set $\mathbb{A}_{v_{rel}}$. This may reproduce the reaction of different drivers at the beginning of a critical braking manoeuver.

Let us denote the probability density over $\mathbb{A}_{v_{rel}}$ by $p_a(a_{rel}|v_{rel})$ For. each $a_{rel} \in \mathbb{A}_{v_{rel}}$ we compute whether a collision will happen (i.e. t_c exists and $a_{rel} \in \mathbb{A}_c$) or not $(a_{rel} \in \mathbb{A}_{nc})$. Obviously, $\mathbb{A}_{v_{rel}} = \mathbb{A}_c \cup \mathbb{A}_{nc}$.

For all $a_{rel} \in \mathbb{A}_c$ we can consider their corresponding t_c. The distribution $p_t(t_c)$ of these values can be derived by

$$p_t(t_c) = p_a(a_{rel}) \cdot \left.\frac{da}{dt}\right|_{(t_c, x_{rel}, v_{rel})} \tag{4}$$

where $a(t,x,v) = T_c^{-1}(t,x,v)$ and $a_{rel} = a(t_c, x_{rel}, v_{rel})$, if the derivative of the inverse of the function T_c of the variable t exists at (t_c, x_{rel}, v_{rel}). Due to the fact that usually not

CHAPTER 7

every value of a_{rel} leads to a collision in positive time and therefore to a physical t_c, we can see that $p_t(t_c)$ is usually not a probability density due to

$$\int_0^{\infty} p_t(t_c)dt_c < 1. \tag{5}$$

We call $p_t(t_c)$ an incomplete density function. The missing part denotes the probability of no crash occurring, i. e.

$$P_{nc} = 1 - \int_0^{\infty} p_t(t_c)dt_c \tag{6}$$

Whereby

$$P_{nc} = \int_{A_{nc}} p(a_{rel})d\mu(\mathbb{A}_{nc}) \tag{7}$$

$$\mathbb{A}_{nc} = \{a_{rel} | t_c \nexists\} \tag{8}$$

We now have a direct measure whether a collision can still be averted or not. If $P_{nc} = 0$, every initial trajectory sampled over a_{rel} according to the density $p_a(a_{rel})$ leads to a collision and the collision is therefore unavoidable. The maximum and minimum time remaining until the collision can be computed using

$$\begin{aligned} t_{max,c} &= \inf\left\{T \geq 0 | \int_0^T p_t(t_c)dt_c = 1\right\} \\ t_{min,c} &= \sup\left\{T \geq 0 | \int_0^T p_t(t_c)dt_c = 0\right\} \end{aligned} \tag{9}$$

We now discuss how a given situation can evolve in space. Let us therefore consider a set $\{s_n \in S | n \in \mathrm{N}\}$ of scenarios in which we reduce the relative distance x_{rel} between the two objects; let v_{rel} be constant and sample a_{rel} according to a probability density $p_a(a_{rel})$. Each of the scenarios s_n has its own (incomplete) density function $p_{n,t}(t_c)$ and probability of no crash $P_{n,nc}$. The scenario s_N in which for the first time $P_{N,nc} = 0$ is labeled the scenario of unavoidability. For this scenario we define the time of unavoidability by

$$t_{ua} = t_{max,c} = \inf\left\{T \geq 0 | \int_0^T p_{N,t}(t_c)dt_c = 1\right\} \tag{10}$$

This quantity is a crucial parameter when developing algorithms because it indicates that no further possible scenario can avoid the collision.

Let us now have a look at some examples. We start with an initial example with x_{rel} = 10 m, v_{rel} = 5 m/s and a_{rel} is sampled according to a Gaussian distribution with parameters $\mu = 0$ m/s^2 and $\sigma^2 = 1$ (m/s^2)^2. Based on these initial conditions, we can directly derive the incomplete density distribution $p_t(t_c)$ which is depicted in Figure 1. Note that this distribution is not symmetric despite the symmetric initial values and that the maximum of this distribution does not coincide with the t_c value which corresponds to the most probable acceleration $a_{rel} = 0$ m/s^2.

If we now generate a sequence of scenarios where we reduce the initial distance x_{rel} from 10 m to 0 m we get the following evolution of the (incomplete) time to collision probability density (Figures 2 and 3). Figure 2 shows a plane view of the evolution and Figure 3 the full three dimensional view. As expected the maximum of the distribution and its width approaches zero with reduced distance to target. Furthermore, the integral of the incomplete probability distribution approaches one. This is due to the fact that more and more possible acceleration values are not able to prevent the crash from happening.

The probability of no crash evolves accordingly (see Figure 4), i.e. at long distances the probability is nearly one. It approaches zero while reducing the initial distance. Note that there is a relatively small area in which this change takes place.

FIGURE 1 Probability density of time of collision for starting scenarios which are sampled according to a Gaussian distribution in acceleration space.

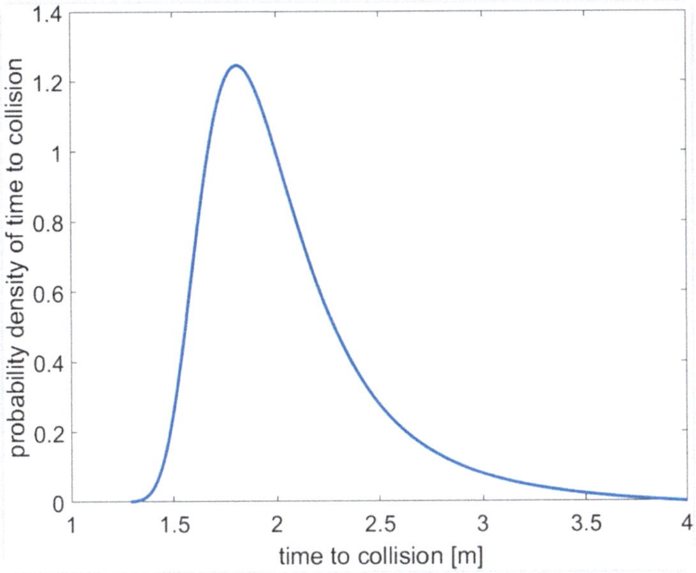

FIGURE 2 Spatial evolution of the probability density of time of collision (plan view).

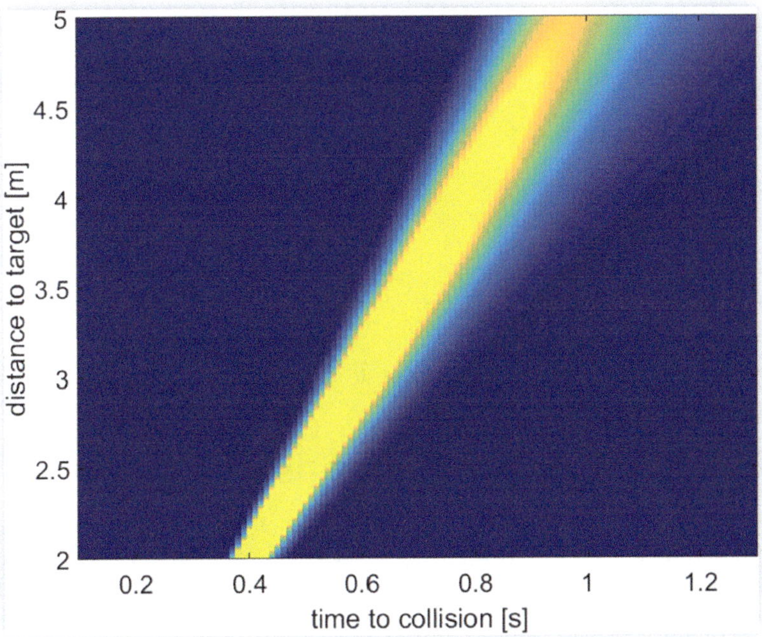

FIGURE 3 Spatial evolution of the probability density of time of collision (full view). Note the reduction of the variance while approaching the target.

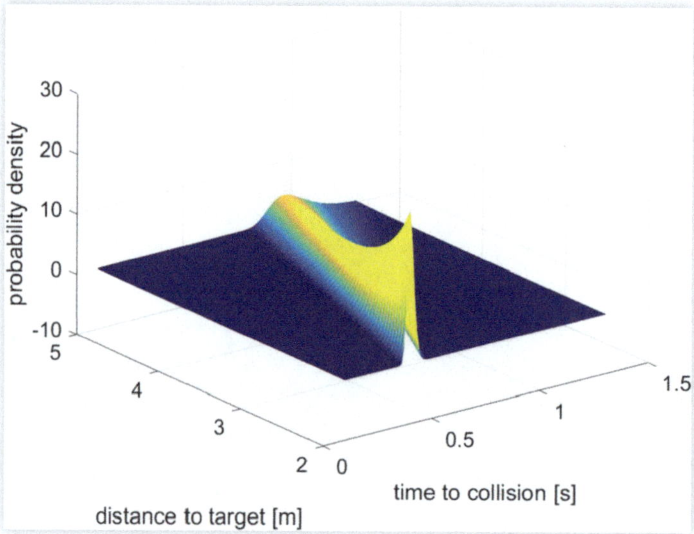

FIGURE 4 Spatial evolution of the probability of no crash.

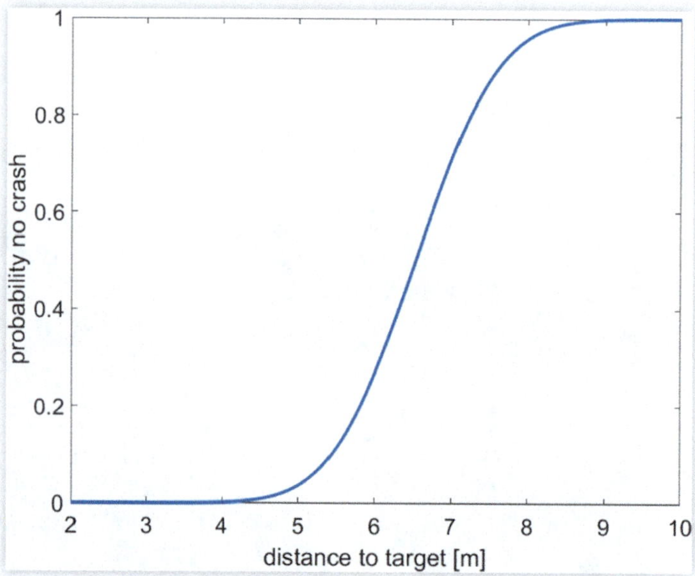

Moving to Two Dimensions

Let us now have a look at the two dimensional dynamics. The first example is generalized to two dimensions by introducing the second control parameter a_y. Furthermore, the relative coordinates have to be skipped because of the connection between longitudinal and lateral dynamics. We also consider the target width. We still hold on to the no change assumption after defining/sampling the initial scenario parameters.

Let

$$S = \left\{ \left(x^n, v^n, a^n \right) | x^n \in \mathbb{R}^2, v^n \in \mathbb{V}^2, a^n \in \mathbb{A}_v^2, n = 0,1 \right\} \tag{11}$$

FIGURE 5 Numeric approximation for the probability density of time to collision. For both vehicles the lateral and longitudinal acceleration was sampled used Gaussian distribution with μ = 0 m/s^2 and σ^2 = 2 $(m/s^2)^2$ was used.

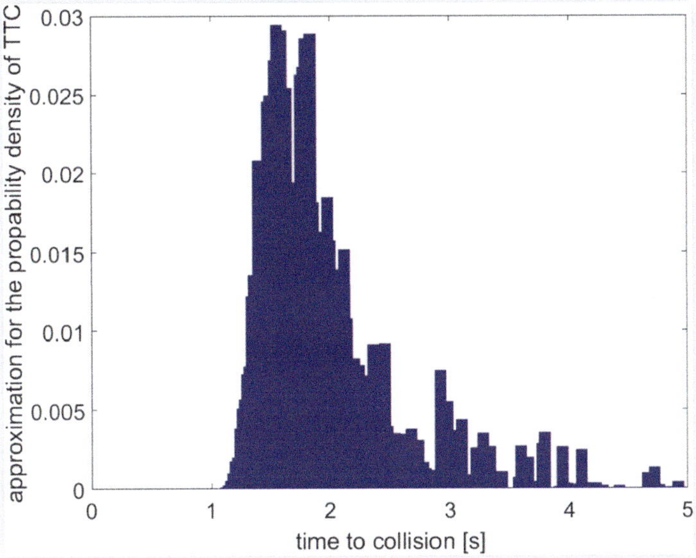

be the set of possible scenarios. The superscript n denotes the considered vehicle. Using the no change assumption we can again introduce the time to collision t_c, if it exists for a given scenario and discuss its incomplete density distribution which is related to an initial acceleration sampling. The probability for no collision can be computed accordingly.

But this problem is very hard to solve analytically, especially because of the vehicle dimensions that have to be considered in order to detect collisions. Because of that, it is more practical to generate a numerical approximation. This can be achieved by a simulation approach, where the lateral and longitudinal vehicle dynamics is based on a sufficient complex vehicle model. With that, a high resolution sampling of several possible accelerations for a given scenario has to be simulated.

As an analogy to the 1D example described in the previous section, we constructed a similar scenario in two dimensions. The first vehicle starts at $x^0 = (0 \text{ m}, 0 \text{ m})^T$ with a velocity of $v^0 = (20 \text{ m/s}, 0 \text{ m/s})^T$. The second vehicle starts at $x^1 = (15 \text{ m}, 0 \text{ m})^T$ with a velocity of $v^1 = (15 \text{ m/s}, 0 \text{ m/s})^T$. With this parameters (and the corresponding vehicle lengths) the 2D scenario reflects the 1D scenario with $x_{rel} = 10$ m and $v_{rel} = 5$ m/s.

The accelerations (longitudinal and lateral) of both vehicles are sampled according to a Gaussian distribution with parameters $\mu = 0$ m/s^2 and $\sigma^2 = 2$ (m/s^2)^2. In total we simulated 16,000 different variations leading to collision using different acceleration values. Figure 5 shows the distribution of the resulting time to collision. Comparing the result of the sampled two dimensional simulation with the analytic solution of the one dimensional scenario (see Figure 1) shows the similarities of both approaches and indicates, that we can approximate the correct solution with simulation of sampled data.

The graph in Figure 6 shows another result of such a simulative approach to obtain the probability distribution of no crash for a scenario $s \in S$. However, the generation of such data is computationally very intensive, hence time consuming: the generation of the data shown in Figure 6 took almost eight hours on a standard workstation.

FIGURE 6 Numerically obtained collision probability. The value 1 is reached at a time to collision of 0.060 s.

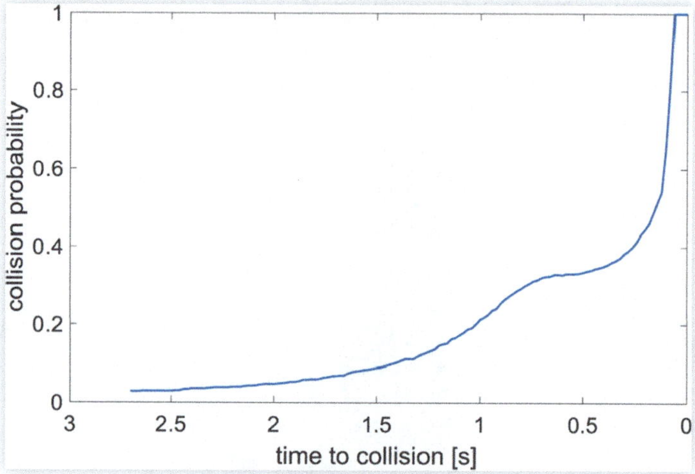

Limiting Trajectories and Sampling in between

We now want to discuss how we can reduce the computational effort without losing most of the results obtained before. To do this we replace the computationally costly high resolution sampling by a subsampling which takes only into account a small number of acceleration values. This can be described by introducing the point measure

$$p_a(a) = \frac{1}{N}\delta_{\mathbb{A}_p}(a), \tag{12}$$

Where $\mathbb{A}_p$ is the set of the acceleration values considered and $N = |\mathbb{A}_p|$ is the number of elements of this set.

To consider limiting scenarios we use the set

$$\begin{aligned} \mathbb{A}_p = \{(a_x^n, a_y^n) | (a_x^n, a_y^n) = K(\hat{a}_x^n, \hat{a}_y^n | R), \\ \hat{a}_x^n \in \{a_{x,min}^n, 0, a_{x,max}^n\}, \\ \hat{a}_y^n \in \{a_{y,min}^n, 0, a_{y,max}^n\}, \\ n = 0,1\} \end{aligned} \tag{13}$$

where $a^n_{(x,y),(min,max)}$ are the boundary values for the lateral and longitudinal accelerations. The resulting combinations of accelerations are limited by Kamm's circle $K(\hat{a}_x, \hat{a}_y | R)$ with radius R to prevent the generation of unrealistic scenarios.

The final analysis then reduces to the examination of the behavior of the N resulting trajectories. Since equation (7) reduces to

$$P_{nc} = \frac{N_{A_{nc}}}{N} \tag{14}$$

where N_{Anc} denotes the number of trajectories given by the set $\mathbb{A}_p$, which do not lead to a collision, the probability for no collision can be obtained directly by computing a simple fraction. The result, however, is just an approximation of the outcome that would be achieved by considering the entire possible dynamics.

To determine the value of N_{Anc} for a given scenario $s \in S$, any sufficient vehicle dynamics simulation tool can be used. In the first step the scenario s is simulated with a fixed simulation time step to determine the time of collision t_{co}. Next, a second simulation procedure is performed, where the scenario is re-simulated at every time step prior to t_{co}. The re-simulation thereby consists of N individual simulations, where the accelerations of the considered vehicles are altered according to the elements of $\mathbb{A}_p$. This results in a value P_{nc} for every time step and altogether in the probability distribution for no collision over time (cf. Figure 4).

TABLE 1 Parameters used for the estimation of t_{ua} by simulation

Parameter	Value used (unit: m/s²)
$a^0_{x,min}$	-9
$a^0_{x,max}$	3
$a^0_{y,min}$	-9
$a^0_{y,min}$	9

Note, that if just the value of t_{ua} is of interest and not the complete probability distribution, the re-simulation can be aborted as soon as a trajectory leads to the avoidance of the collision.

To demonstrate the results on an example, we consider a scenario where the ego vehicle collides with a static box shaped object (see also Figure 8; the detailed description of this scenario is not relevant for now, but follows later). We assume the maximal possible accelerations for the ego vehicle ($n = 0$) as shown in Table 1.

Since the second object ($n = 1$) is static, all possible accelerations $a^1_{(x,y),(min,max)}$ can be set to zero. Thus the set $\mathbb{A}_p$ can be reduced to a total of 9 unique elements. Furthermore a radius of 9 m/s² for Kamm's circle is used.

As shown in Figure 7, the plot of the values $P_c = 1 - P_{nc}$ shows a step function that represents the unavoidability of the collision for any relevant time to collision.

The value of t_{ua} can now be derived from the graph by picking the time of collision where the value of the step function reaches 1 for the first time. In our example, the time of unavoidability is reached 60 ms before the collision.

Note that the plot looks similar to the no-crash probability shown in Figure 6. That is because the computation method has been used on the same scenario. It can be seen, that the approximation method shown in this paragraph is sufficient to obtain the value of t_{ua} and even the rough probability distribution of no collision itself.

What Are the Problems?

Let us first discuss the following example using a stationary target. In this simple scenario the vehicle is moving towards a static obstacle located at the origin of the world coordinate system with constant steering angle and constant acceleration. The parameters are

FIGURE 7 Portion (P_c) of the maximal ego manoeuvres that will result in a collision with the static object, if performed at the corresponding time to collision.

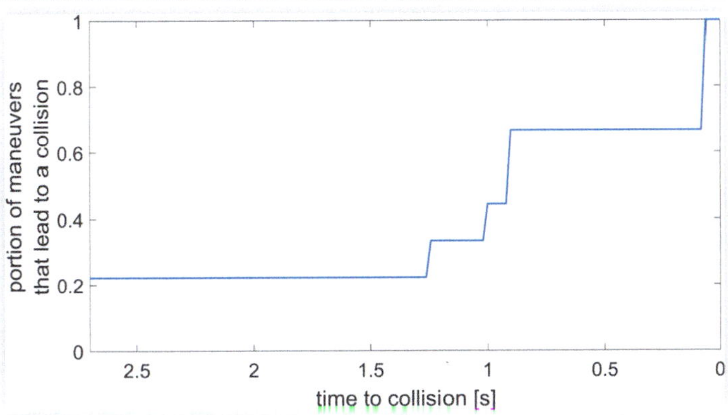

TABLE 2 Scenario parameter

Start-position(X|Y)	−43.0 m|−20.4 m
Start-heading	48.1°
Start-velocity	15.7 m/s
Constant acceleration	0.5 m/s^2
Constant steer angle	−2.9

given in Table 2. Without any additional action the vehicle collides at time t =2.7 s. Figure 8 illustrates this simple scenario.

Let us assume the vehicle is equipped with a perfectly working automatic emergency braking system, which operates without any latencies.

In the given scenario (Figure 8) the emergency braking system has to apply full braking 900 ms before the collision occurs (marked with the orange X), otherwise the collision cannot be prevented by the emergency braking system. The driver on the other hand can still prevent the accident with applying a gentle braking and steering starting 300 ms before the collision (marked with the purple pentagon). This time difference is caused by the different actions the system and the driver can use and makes it impossible to implement an emergency braking system avoiding all collisions by braking while having zero false alarms at the same time. This kind of requirement conflict exists basically in every driver assistance system without full control over the vehicle's actuators. One possible work-around is to change the system's goal to only minimize the relative velocity of the collision instead of preventing it. The other implication would be allowing false alarms, which normally decreases the customer's satisfaction.

Figure 9 shows another scenario, where two vehicles are driving on a gentle curve on oncoming lanes. The vehicle of interest, called ego, is driving a gentle left curve beginning at the south-east corner. The oncoming vehicle (target) is driving towards the ego vehicle. Both vehicles are driving with a velocity of 80 km/h and are therefore representing a typical inter-city, non-highway situation.

Applying numerous variations on the start position, velocity and steering angles of the two vehicles results in many scenarios including collisions. To determine the absolute minimum field of view to detect the target vehicle before the collision happens, we take a look at the location of the target vehicle from the ego's point of view at the time of unavoidability. Theses target positions at the time of unavoidability are shown in Figure 10. The data gives us a first indication of the areas we have to look at for designing

FIGURE 8 Simple example scenario, showing the vehicle location at the time of collision (t = 2.7 s). The trajectory without any change in steering is drawn in blue. The orange X marks the latest point, where the collision can be avoided by applying braking only. The purple trajectory shows an alternative trajectory avoiding the collision by using gentle steering and braking.

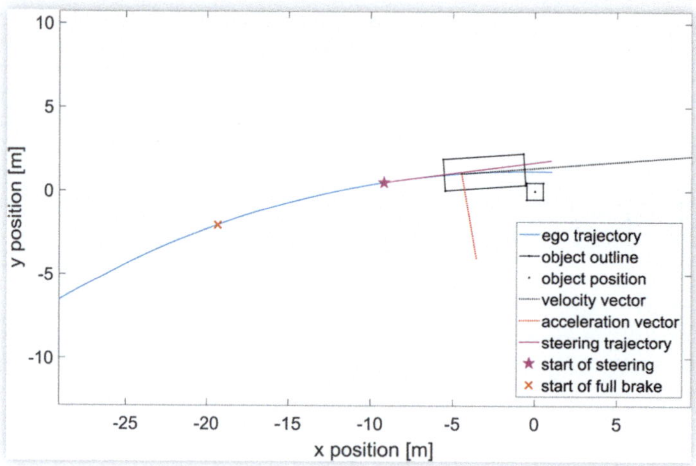

FIGURE 9 Base scenario without collision. Ego is driving a gentle left curve, whereas the oncoming traffic is on the opposite lane.

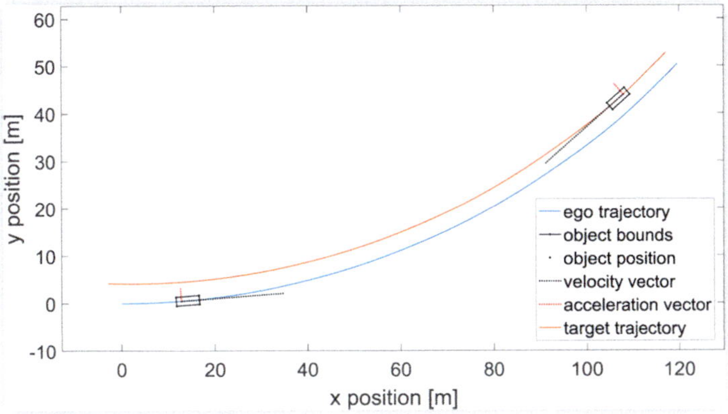

FIGURE 10 Position of target vehicle at t_{UA} and an arbitrary sensor cone with an aperture angle of 45° and a range of 25 m.

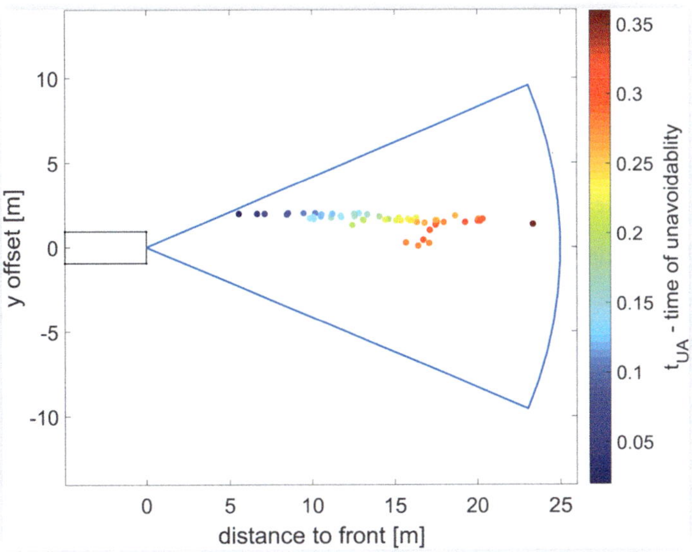

a collision assistance function. Note that we are still assuming, that we have a perfect sensor, which only needs one measurement for detecting an object.

The data depicted in Figure 10 shows only the location where the collision is physically unavoidable, meaning even if both vehicles make use of their maximal longitudinal and lateral acceleration potential, the collision will still occur. After this point in time the driver assistance system can only try to minimize the impact of the collision.

The plot in Figure 10 also suggest, that a perfect sensor with an aperture angle of 45° and a detection range of 25 m is sufficient to detect the target vehicle before the collision is unavoidable. This observation is only the first clue for the required sensor field of view.

FIGURE 11 Showing the target's position (from the ego's point-of-view) at the times between t_{UA} and $t_{latency}$ = 400 *ms* before t_{UA}. The blue sensor cone is the same as shown in Figure 10. The orange sensor cone has an aperture angle of 22.5° and a range of 45 m.

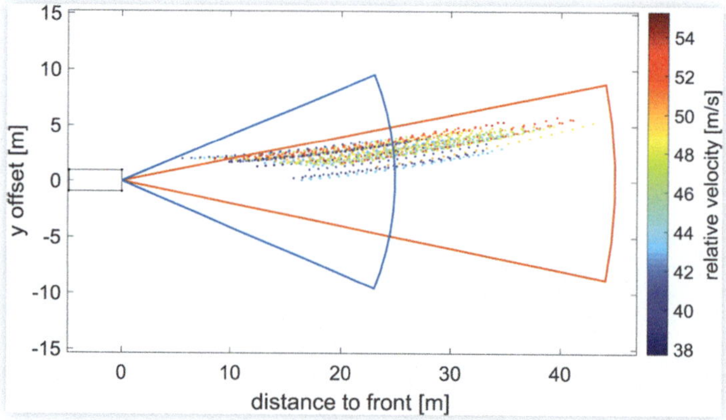

With this curve scenario we can show another conflict for pre-crash system, so let us assume the following:

- both vehicles are equipped with a similar crash avoidance system
- these systems are allowed to use the full range of braking/ acceleration and steering of the vehicle
- the sensor sets of both vehicles work perfectly (no false measurements, no uncertainty)
- the drivers are not overriding the system

These assumptions are very strong and never met in reality, but they allow us to discuss the limitations of such a system by only introducing latency. For this example we define the latency $t_{latency}$ of the whole system (including sensor data processing, the decision algorithm itself, the delay time of the actuators, etc.) to be 400 ms, which is a realistic value for real-world systems.

Due to the latency, the target has to be detected 400 ms before the time of unavoidability to react in time. The impact of this latency in terms of sensor field of view is shown in Figure 11. As first implication the sensor range has to be increased by at least 20 m (or 80%) in return the aperture angle can be reduced to 22.5° (blue sensor cone). Although it appears to miss a few positions with the tighter sensor cone, these positions do not matter, because they are detected at least $t_{latency}$ before the time of unavoidability t_{ua} and therefore in time for the decision to apply the emergency braking.

Another issue caused by the latency is, that the decision of the system's intervention has to be done before the collision is unavoidable. This means the system has to react at a point in time, when the collision is still avoidable by the driver, which in turn leads to false-positives by force. Alternatively if the system reacts not until the time of unavoidability t_{ua}, the effect of the system would be reduced. Depending on the system's actuator, e.g. an external structure-hardening airbag, it may even be rendered useless.

Now we discuss yet another scenario, in the following example (shown in Figure 12) the two vehicles, ego and target, driving on a curve are avoiding a collision by only a few centimeters. Both vehicles are driving with constant acceleration and steer angle.

FIGURE 12 Ego and target vehicle do not collide, but they miss each other by only a few centimeters. This scenario uses no change-assumptions for both vehicles. The image was generated at simulation time t = 2.76 s.

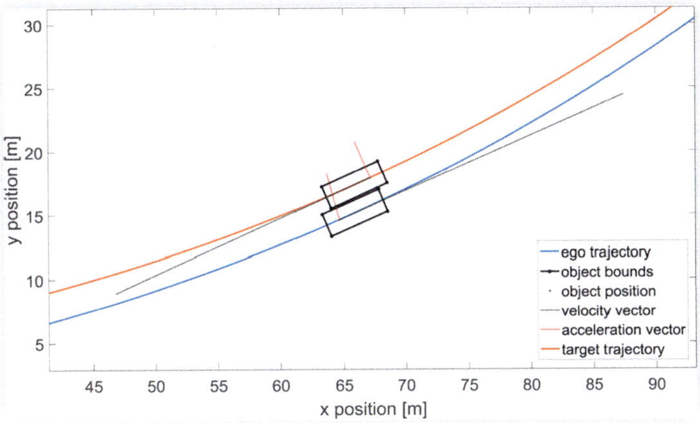

As there is no collision, there is also no time of unavoidability t_{ua} and no need for the collision avoidance function to interfere.

This example shows that very precise sensors are required to be able to differentiate between collision and non-collision, or near-collision, scenarios. To compensate uncertainty in the measurement the specification of the collision avoidance function has to allow false-negatives (not reacting although a collision is going to happen) or treat near-collision situation as collision and react accordingly.

There is another problem, especially but not limited to near-collision scenarios, which is the unpredictability of human drivers. A small change in the driving maneuver (steering, braking/accelerating) caused by the human driver leads to a practical unpredictability of the vehicle's trajectory and therefore also of the collision or time of unavoidability.

What Can Be Done

The discussion above has shown that current systems will not be able to fully eliminate critical frontal collisions between two vehicles due to incomplete information about their future trajectories.

One of the main problems hereby is the behavior of the driver - because his future reactions are uncertain, the algorithms are not able to predict whether a collision will happen or not. Normally, we can only compute some kind of collision probability which evolves depending on the future development of the scenario.

But even if we eliminate the driver from the discussion and consider only self-driving or autonomous vehicles, it is clear that due to the restriction of the sensors in the vehicle some of the conflicting scenarios described above can still not be solved and accidents cannot be avoided. So what can be done to eliminate these problems? The easiest way to avoid such accident scenarios is to give the vehicles and therefore the algorithms more information about the future development. This information can be shared using Car2Car-communication systems which can transmit information about the desired trajectory of the vehicles. Another approach is to centralize the trajectory planning of the autonomous vehicles which ensures that the planned trajectories will not interfere with each other. This will at least circumvent the problems described above for self-driving vehicles.

But as long as a human being is in the loop, we surmise that accidents will happen.

Summary/Conclusions

In this paper we have shown the fundamental limitations of current forward collision avoidance systems. These limitations are mainly defined by the incomplete information that the systems have when crucial decisions have to be done. The shortages of sensors, system latencies, the stochastic nature of traffic scenarios, the unknown behavior of the driver, the high dynamics when considering oncoming traffic all contribute to this lack of information. So these systems will help to reduce fatalities and injuries, but will not be able to prevent all accidents. Other measures like sharing information using Car2Car communications or a centralized path planning for autonomous cars have also to be taken into account to reach Vision Zero.

Contact Information

The authors can be contacted using the following address:
IEE Sensing Germany, **Dr. Frank Keck**
Ruppertswies 14, 85092 Kösching, Germany
frank.keck@iee.lu

Definitions/Abbreviations

Kamm's circle - circle of forces
latency - time delay due to computational, sensorial, ... influences
limiting trajectories - maximal possible trajectories
time of unavoidability - time after which accident is no longer avoidable
time to collision - time until accident

References

1. *Global Status Report on Road Safety 2015* (World Health Organization), ISBN 978 92 4 156506 6.
2. "Unfallentwicklung auf deutschen Straßen 2015," Statistisches Bundesamt, http://www.destatis.de/publikationen.
3. Anderson, R.W.G., Doecke, S.D., Mackenzie, J.R.R., and Ponte, G. et al., "Potential Benefits of Forward Collision Avoidance Technology," Centre for Automotive Research, Report Number CASR0106.
4. Jansson, J., "Collision Avoidance Theory," Linköping Studies in Science and Technology, Dissertations No. 950.
5. Althoff, M., "Reachability Analysis and its Application to the Safety Assessment of Autonomous Cars," Technische Universtität München, 2010.
6. Althoff, M., Stursberg, O., and Buss, M., "Model-Based Probabilistic Collision Detection in Autonomous Driving," *IEEE Transactions on Intelligent Transportation Systems* 10, no. 2 (June 2009).
7. Broadhurst, A., Baker, S., and Kanade, T., "Monte Carlo Road Safety Reasoning," *Proc. IEEE Intell. Vehicles Symp*, Las Vegas, NV, June 2005, 319-324.

8. Wheeler, T., *Probabilistic Driving Models and Lane Change Prediction*, 2014.
9. Zhao, D., Lam, H., Peng, H., Bao S. et al., "Accelerated Evaluation of Automated Vehicles Safety in Lane Change Scenarios based on Importance Sampling Techniques," arXiv:1605.04965.
10. Houénou, A., Bonnifait, P., and Cherfaoui, V., "Risk Assessment for Collision Avoidance Systems," *17th International IEEE Conference on Intelligent Transportation Systems (ITSC)*, Qingdao, 2014, 386-391, doi:10.1109/ITSC.2014.6957721.
11. Laugier, C., Paromtchik, I., Perrollaz, M., Mao, Y. et al., "Probabilistic Analysis of Dynamic Scenes and Collision Risk Assessment to Improve Driving Safety," *Its Journal, Informa UK (Taylor & Francis)* 3, no. 4 (2011): 4-19, doi:10.1109/MITS.2011.942779.
12. Lee, D.N., "A Theory of Visual Control of Braking Based on Information about Time-to-Collision," *Perception* 5, no. 4 (1976): 437-459.

CHAPTER

8

Hardware-in-the-Loop (HIL) Implementation and Validation of SAE Level 2 Autonomous Vehicle with Subsystem Fault Tolerant Fallback Performance for Takeover Scenarios

Adit Joshi
Ford Motor Co., Ltd.

CHAPTER 8

The advancement towards development of autonomy follows either the bottom-up approach of gradually improving and expanding existing Advanced Driver Assist Systems (ADAS) technology where the driver is present in the control loop or the top-down approach of directly developing Autonomous Vehicles (AV) hardware and software using alternative approaches without the driver present in the control loop. Most ADAS systems today fall under the classification of SAE Level 1 which is also referred to as the driver assistance level. The progression from SAE Level 1 to SAE Level 2 or partial automation involves the critical task of merging autonomous lateral control and autonomous longitudinal control such that the tasks of steering and acceleration/deceleration are not required to be handled by the driver under certain conditions [1]. However, the driver is still required to monitor the driving environment and handle scenarios where control is handed over to the driver due to subsystem faults of the autonomous system. Due to the disadvantages of vehicle testing being expensive, time-consuming and hazardous for testing such scenarios, an alternative method of development and validation is required. Therefore, the objectives of this research are two-fold. The first objective focuses on a real-time powertrain-based Hardware-in-the-Loop (HIL) implementation and validation of an SAE Level 2 autonomous vehicle. The second

CITATION: Joshi, A., "Hardware-in-the-Loop (HIL) Implementation and Validation of SAE Level 2 Autonomous Vehicle with Subsystem Fault Tolerant Fallback Performance for Takeover Scenarios," SAE Technical Paper 2017-01-1994, 2017, doi:10.4271/2017-01-1994.

objective focuses on studying the performance of SAE Level 2 autonomous vehicles during takeover scenarios due to subsystem faults. To accomplish these objectives, an acceleration-based Adaptive Cruise Control (ACC) was combined with a path-following lateral control along with supervisory control for system mode transitions due to system deactivations and faults. This research presents system modes in which longitudinal control only and lateral control only are engaged as fallback states to the full autonomous system being faulted for lateral control and longitudinal control failures respectively. Simulations were conducted to evaluate the performance of the autonomous controls when subjected to these faults. A powertrain subsystem representative of the 2017 Ford Fusion Hybrid was used as the hardware simulation platform using a dSPACE HIL simulator and CarSim RT.

Introduction

The development of ADAS and AV technology has been accelerating in recent years due to safety concerns arising from human driver error related crashes. According to NHTSA, the number of fatalities due to automobile accidents in the US in 2015 was approximately 35000, with an additional 2.3 million reported injuries. Approximately 94% of these fatalities and injuries were caused by various forms of human driver error such as distracted driving, impaired driving, speeding, driver inexperience, driver fatigue etc. [2, 3, 4]. According to NHTSA, in 2000, the cost of roadway crashes for the US economy was estimated at $212 billion [5]. Cost benefit research and studies have shown that the maximum potential saving per year using ADAS and AV technology can be approximately $180-190 billion if automobile accidents are reduced by a factor of 90% [6]. Although autonomous vehicles have additional benefits such as optimizing traffic management, improving fuel economy, reducing emissions, improving driving convenience, etc., the main benefit remains that of increased safety of the occupants of the vehicle and other road users such as occupants of other vehicles, pedestrians and bicyclists by reducing automobile related fatalities and injuries [4, 7].

The advancement towards development of autonomous vehicles follows either the bottom-up approach of gradually improving and expanding existing ADAS technology where the driver is present in the control loop or the top-down approach of directly developing AV hardware and software using alternative approaches without the driver present in the control loop. The motivation for the former approach is driven by regulations for active safety of vehicles and a need for increasingly capable semi-autonomous operation from development of new automated driving features. SAE classifies autonomous vehicles broadly into six levels of automation ranging from no automation (SAE Level 0) to full automation (SAE Level 5) with most ADAS systems today falling under the classification of SAE Level 1 which is also referred to as the driver assistance level [1].

The progression from SAE Level 1 to SAE Level 2 or partial automation involves the critical task of merging autonomous lateral control and autonomous longitudinal control such that the tasks of steering and acceleration/deceleration are not required to be handled by the driver under certain conditions [1]. The benefit of such an autonomous vehicle is that it temporarily relieves the driver of the dynamic driving tasks in certain conditions that do not require high cognitive load from the driver such as driving on

TABLE 1 SAE levels of automation [1]

Level	Name	Driving Role	Monitoring Role	Fallback Role
0	No Automation	Human Driver	Human Driver	Human Driver
1	Driver Assistance	Human Driver & System	Human Driver	Human Driver
2	Partial Automation	System	Human Driver	Human Driver
3	Conditional Automation	System	System	Human Driver
4	High Automation	System	System	System
5	Full Automation	System	System	System

highways with little or no traffic. An SAE Level 2 autonomous vehicle accomplishes this by taking over longitudinal control in terms of acceleration and braking and lateral control in terms of steering. However, the driver is still required to monitor the driving environment and is responsible for the fallback performance of the dynamic driving tasks. The human driver is responsible for handling scenarios where control is handed over to the driver in situations where the operational envelope of the automated system is exceeded. The human driver is also responsible for intervening and taking control of the vehicle if required, since the system cannot recognize its performance envelope and limits. The SAE classification of the levels of automation showcasing the different roles of the human driver and system is shown in Table 1 [1].

An SAE Level 2 autonomous vehicle involves automation of at least two primary control functions designed to work in unison to relieve the driver of control of those functions. A key feature of SAE Level 2 automation or partial automation is that it involves the capability of fully controlling the vehicle for limited periods in restricted situations [8]. The most common application of an SAE Level 2 autonomous vehicle is accomplished by combining adaptive cruise control with path following or lane centering. More than 50% of all fatal accidents in 2009 in the U.S. occurred due to some form of aggressive driving such as over-speeding, erratic maneuvering, improper or illegal driving, etc. [9]. Other forms of aggressive driving behaviors include tailgating, weaving through traffic, violating traffic laws and violating traffic control devices. Aggressive driving is defined as a deliberate driving behavior which increases the risk of collision and is motivated by impatience, annoyance, hostility, and/or an attempt to save time by the human driver [10]. One of the potential benefits of this type of autonomous vehicle would be the mitigation of a human driver's intentions and attempts for aggressive driving and maneuvering, such that risks of potential hazardous driving situations are reduced [8].

For automating a system, there are two types of automation. The first type of automation is event-based automation in which there is continuous monitoring but automation only takes control for a short period of time in response to a specific event. The second and more robust type of automation is continuous automation in which not only is there continuous monitoring but also automation continuously controls the system for extended periods of time [11]. Continuous automation may be deactivated due to a decision made by the automated controls system or due to input from the driver indicative of driver desire to take over vehicle controls by actions such as depressing the brake pedal or applying torque to the steering wheel. Continuous automation may also be deactivated due to the detection of a fault or failure of the automated control system to generate the correct control commands [12]. Such situations provide justification for the need for driver takeover. Since the automated control system will fail to generate proper commands to the actuators due to failures or faults, the driver may still be able to take over control of the vehicle by commanding the actuators in such situations where the capability of the driver to control the vehicle remains. [11].

The high complexity of hardware and software required for an autonomous vehicle also leads to numerous potential sources of automation failure [11]. An automation failure occurs when automation no longer delivers proper performance [13]. All components of the automation system are potential sources of failure. When a component or a complete system fails, its performance deviates from the intended correct performance such that the resulting performance may be poor or degraded [13]. There are varying degrees of failure modes and fault severity. These range from minor faults at the component level such as signal misalignments and offsets to major faults which result in complete subsystem or system failures [11, 14]. This paper only considers relatively minor faults such that the vehicle drivability is maintained even with degraded vehicle performance [14].

Due to the disadvantages of vehicle testing being expensive, time-consuming and hazardous for testing such scenarios, an alternative method of development and validation is required. Moreover, software required for controls of autonomous vehicles is highly complex, using in-vehicle testing alone will require vast amount of testing by accumulating driven kilometers to ensure complete validation of the software. According to the RAND Corporation, a 100 vehicle fleet running 24 hours a day 365 days a year at a speed of 40 km/hr, would require 17 billion driven kilometers of testing, which in turn would require 518 years to fully validate the software with 95% confidence such that the software failure rate would be 20% better than the current human driver fatality rate [15]. Therefore there is a realistic need for alternative methods of testing that can accelerate and shorten the development and validation of autonomous vehicles. HIL testing a method that can address that need by providing a simulation platform in which rapid prototyping and testing of control software and hardware can be done in a controlled and consistent testing environment for automated driving feature development. Therefore, the objectives of this research are two-fold. The first objective focuses on a real-time powertrain-based HIL implementation and validation of an SAE Level 2 autonomous vehicle. The second objective focuses on studying the performance of SAE Level 2 autonomous vehicles during takeover scenarios caused due to subsystem faults. To accomplish these objectives, an acceleration-based adaptive cruise control was combined with a path-following lateral control along with supervisory control for system mode transitions due to system deactivations and failures. In particular, this research includes system modes the autonomous system can enter into in addition to the traditional state of complete autonomous system disengagement. This research presents system modes in which longitudinal control only and lateral control only are engaged as fallback states to the full autonomous system being faulted for lateral control and longitudinal control failures respectively. Simulations were conducted to evaluate the performance of the autonomous controls when subjected to these faults. A powertrain subsystem comprising controllers and actuators representative of the 2017 Ford Fusion Hybrid was used as the hardware simulation platform using a dSPACE HIL simulator and CarSim RT.

Hardware-in-the-Loop (HIL) Simulation

Hardware-in-the-Loop (HIL) testing is a simulation method used for controls systems testing in a real-time simulation environment [16, 17]. It involves the combination of actual controller hardware in the system loop with a simulated process. In a HIL simulation the physical part of the system or the plant is replaced by a plant model simulation which can execute in real-time such that it exhibits the same behavior as the real physical process [16, 17, 18]. The plant model simulation is typically connected to the Electronic Control Units (ECUs) through a HIL simulator which provides similar electrical interfaces of physical signals of the real system processes and dynamics [16, 17]. A diagram of showcasing the different parts of a HIL simulation is shown in Figure 1.

A comparison between vehicle, real-time simulation and HIL simulation is shown in Figure 2.

FIGURE 1 Different parts of HIL simulation.

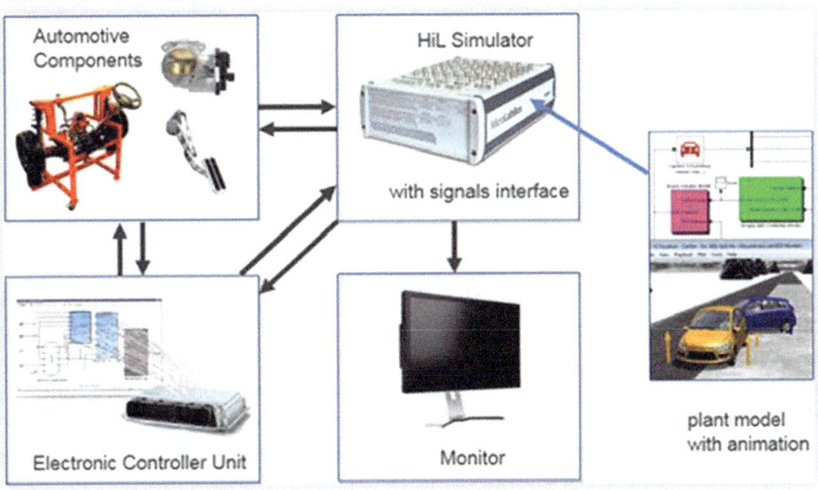

FIGURE 2 Comparison between vehicle, real-time simulation, and HIL simulation [16, 17].

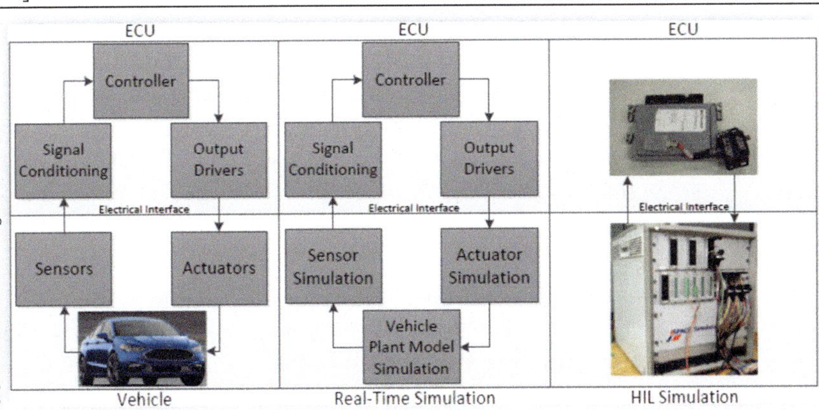

FIGURE 3 Systems engineering V-Cycle [16].

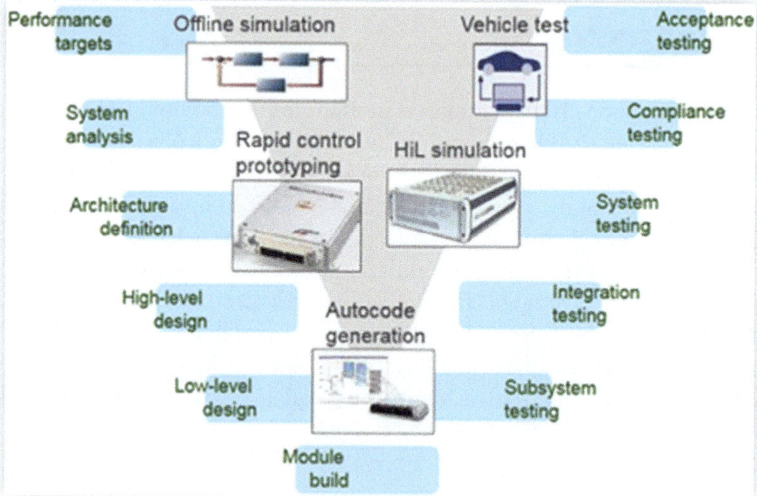

As HIL testing allows for accelerated development and validation of a product, it has become integral for controls and system level engineering and testing. The role of HIL testing in the V-cycles of systems engineering, controls testing and system level testing is shown in Figure 3, Figure 4, and Figure 5 respectively.

HIL Hardware

A dSPACE HIL simulator was the real-time simulation platform used for the HIL simulation by acting as the interface between the hardware components and the model simulation. The dSPACE HIL simulator executed the models in real-time and acted as an emulation of the electrical interfaces to the ECUs. It could support different types of voltage-based physical signals such as Analog, Digital, Pulse Width Modulated (PWM), Resistive, and, Controller Area Network (CAN) signals [17]. The HIL simulation developed and utilized for this research was a powertrain subsystem representation of the 2017 Ford Fusion Hybrid.

The powertrain hardware controllers of the 2017 Ford Fusion Hybrid used for powertrain simulation were Engine Control Module (ECM), Hybrid Control Module (HCU). Moreover, powertrain actuators such as the throttle body, and spark plugs/coils were also included as part of the powertrain HIL setup. Additionally, a Gear Shift Module (GSM) and a Transmission Range Control Module (TRCM) constituting the gear shift-by-wire subsystem were also included in the powertrain hardware setup. To enable the routing of CAN messages between CAN buses, a Gateway Module (GWM) was also added to the HIL setup. Lastly, a Microautobox (MABX) was also included in the setup for rapid prototyping the control features for the automated lateral and longitudinal controls for an SAE Level 2 automated system. A diagram of the hardware setup is shown in Figure 6.

FIGURE 4 Controls development and validation V-Cycle [17, 19, 20].

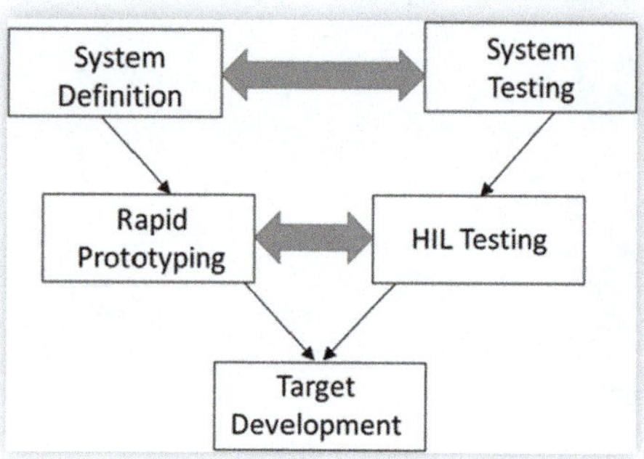

ENGINE CONTROL MODULE (ECM)

This module performs the functions for engine controls in a Full Hybrid Electric Vehicle (FHEV). It enables the translation of the requested powertrain demand into an engine speed and torque commands to the engine. In this research the ECM in combination

FIGURE 5 System development and validation V-Cycle [20].

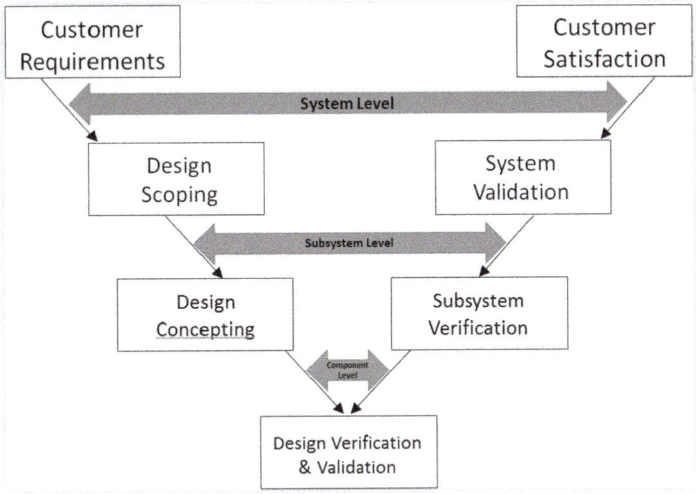

FIGURE 6 HIL hardware setup [17].

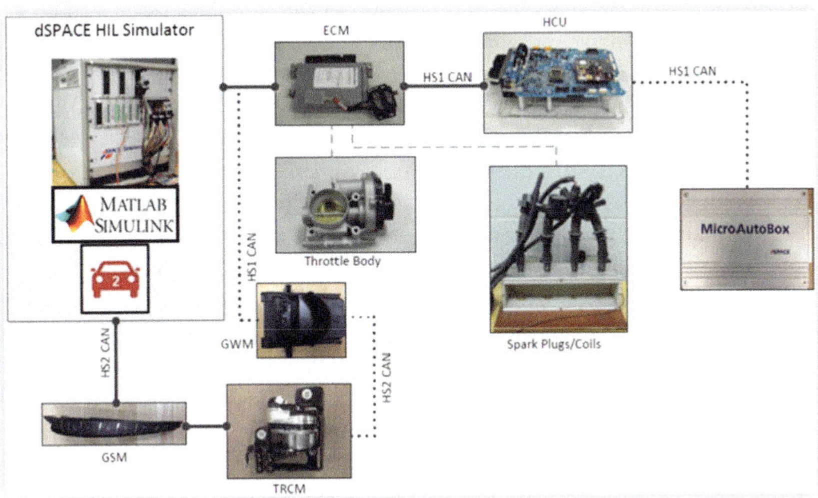

with the HCU was used to actuate the powertrain subsystem in response to commands from the automated longitudinal control in the HIL simulation.

HYBRID CONTROL MODULE (HCU)

This module performs the functions for motor and generator controls in an FHEV. It enables the translation of the requested powertrain demand into generator and motor speed and torque commands for actuating the generator and motor respectively. In this research the HCU in combination with the ECM was used to actuate the powertrain subsystem in response to commands from the automated longitudinal control in the HIL simulation.

GEAR SHIFT MODULE (GSM)

This module consists of a set of push-buttons that allow for selecting Park, Reverse, Neutral, Drive, and Low ranges. It enables the human driver to demand a gear position request and transmits the request to the TRCM.

TRANSMISSION RANGE CONTROL MODULE (TRCM)

This module consists of a motor which is responsible for actuation of the transmission based on desired gear position input from the GSM. It also provides sensor position feedback to the ECM using internal sensors to validate actual gear position.

GATEWAY MODULE (GWM)

This module allows communication and data transfer between the various ECUs located on different CAN buses forming the vehicle network system.

MICROAUTOBOX (MABX)

The MABX is a real-time system which functions as an ECU by executing a Simulink model. The MABX for this research was used for rapid prototyping of the automated lateral control and automated longitudinal control. The MABX was able to transmit the corresponding commands to the HIL simulation through a CAN interface [21, 22].

HIL Modeling

The electrified powertrain using the power-split hybrid architecture is a key feature of the 2017 Ford Fusion Hybrid. The propulsion system for this architecture is comprised of an engine drive system in which the engine, drivetrain and generator are mechanically coupled on a planetary gear set driveline while the electric drive system consists of a high voltage battery and a traction motor. A diagram of the power-split hybrid architecture is showcasing the different aspects of the hybrid powertrain is shown in Figure 7.

The non-hardware components of the 2017 Ford Fusion Hybrid were simulated using a real-time CarSim-Simulink co-simulation environment. A high fidelity plant model of the power-split powertrain comprising an engine, motor-generator, high voltage battery, and planetary gear set driveline was defined in Simulink, which formed the basis of the vehicle level powertrain plant model simulation. The Simulink plant model representation also included high voltage battery and auxiliary controller representations as part of the hybrid propulsion architecture. A CarSim representation of the 2017 Ford Fusion was incorporated to simulate the environment and vehicle dynamics. Due to the high fidelity nature of the CarSim chassis subsystems, the brakes and steering plants were simulated in CarSim. A diagram of the vehicle system level HIL plant model with respect to the controls hardware is shown in Figure 8.

A Simulink-CarSim interface was utilized to conduct the co- simulation between Simulink and CarSim [17]. The Simulink-CarSim interface included dynamic signals for

FIGURE 7 Power-split hybrid architecture.

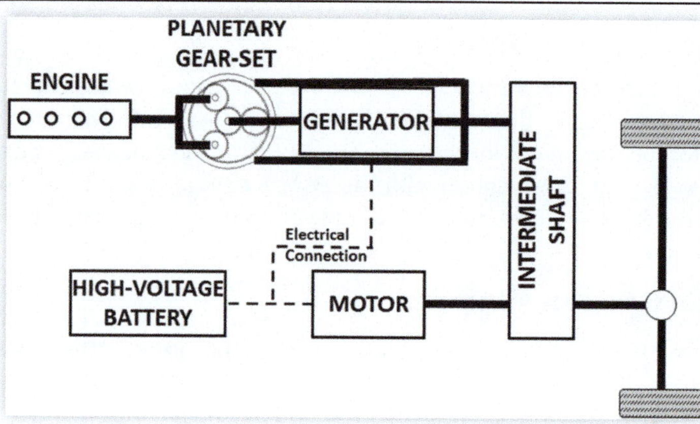

FIGURE 8 Vehicle system level HIL plant model [17].

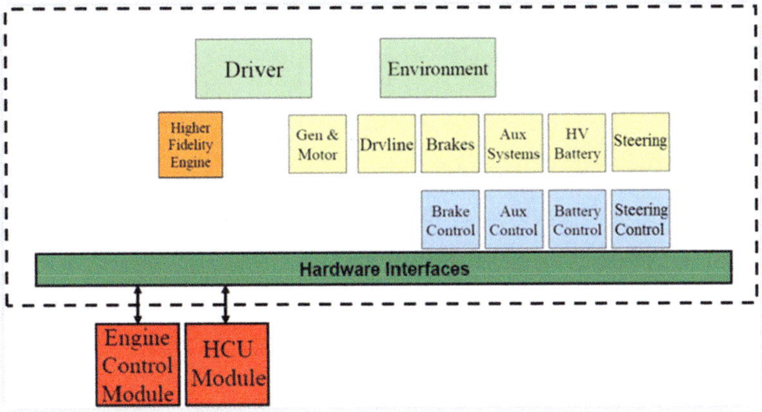

FIGURE 9 Simulink-CarSim interface for vehicle level HIL simulation.

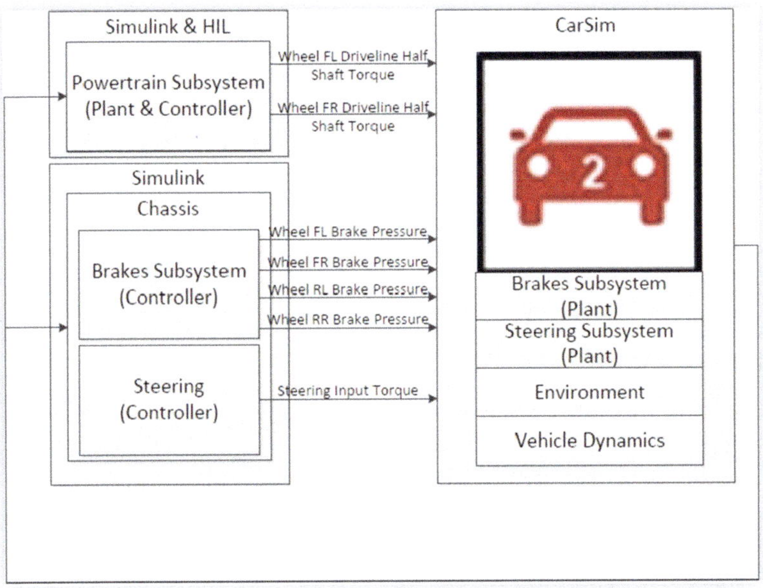

propulsion from the powertrain subsystem, and braking and steering signals from the chassis subsystems. For propulsion action the driveline half-shaft torques from the powertrain plant model were used. The total actual braking torque request from the brakes controller converted to equivalent wheel-end brake pressures represented the braking action.

For steering action, the translated steering wheel torque at the column from the steering angle demand was passed through the Simulink- CarSim interface for actuation of the steering plant present in CarSim. A diagram showcasing the different subsystem controllers and plants with the CarSim-Simulink interface is shown in Figure 9.

Automated Longitudinal Control

For automated driving, similar to non-automated human driving, the primary functions for longitudinal control are throttle and braking [11]. The former function is performed

FIGURE 10 Typical ADAS architecture for control feature.

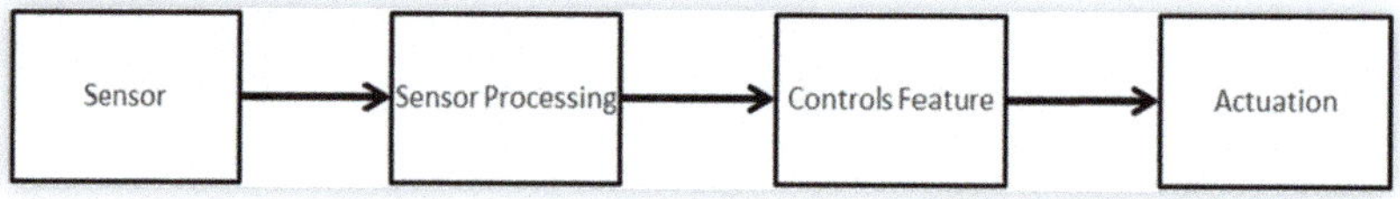

by the powertrain subsystem whereas the latter is performed by the brakes subsystem. For automation of longitudinal control, the objectives are typically to maintain a desired speed, in combination with keeping a desired distance to other vehicles [11].

The longitudinal force (F_x) and force due to powertrain losses (F_{losses}) of the vehicle are functions of the longitudinal acceleration and longitudinal speed respectively as shown in Equations (1)-(2).

$$F_x = m_v a_x \tag{1}$$

$$F_{losses} = F_0 + F_1 v_x + F_2 v_x^2 \tag{2}$$

The factors F_0, F_1, and F_2 in the expression of the force due to powertrain losses are due to longitudinal rolling resistance and aerodynamic resistance. For automated longitudinal control, an acceleration request could be converted into an equivalent torque command using the expressions for the longitudinal force and force due to powertrain losses of the vehicle and tire radius (r_{tire}) as shown in Equations (3)-(4).

$$\tau_x = r_{tire}\left(F_x + F_{losses}\right) \tag{3}$$

$$\tau_x = r_{tire}\left(m_v a_x\right) + r_{tire}\left(F_0 + F_1 v_x + F_2 v_x^2\right) \tag{4}$$

The translated torque command could be decomposed into constituent torques due to powertrain and braking. The constituent powertrain torque ($\tau_{x,powertrain}$) and braking torque ($\tau_{x,brakes}$) are functions of the overall torque (τ_x) and the idle creep torque ($\tau_{x,powertrain,min}$) as shown in Equations (5)-(6)

$$\tau_{x,powertrain} = \max\left(\tau_x, \tau_{x,powertrain,min}\right) \tag{5}$$

$$\tau_{x,brakes} = \tau_x - \tau_{x,powertrain,min} \tag{6}$$

A typical ADAS control feature is able to actuate other downstream subsystems using processed information from a sensor. The architecture of a typical ADAS controls feature is shown in Figure 10.

In this research for the automated longitudinal control, a long-range radar (LRR) was simulated in CarSim. The LRR had two modes, far and near, in which the range and field of view varied. The LRR was placed at the front center of the simulated vehicle with a longitudinal position of 1.150 m from the center of gravity of the vehicle and an elevation of 0.5 m from the base of the vehicle. The sensor aim was pointed at -0.375 deg pitch. The specifications of the LRR are shown in Table 2.

TABLE 2 Specifications of simulated LRR sensor

Specification	LRR Mode 1 (Far)	LRR Mode 2 (Near)
Range	160 m	40 m
Bearing Angles	-10 to 10 deg	-45 to 45 deg
Field of View	20 deg	90 deg

Information about the target vehicle such as distance and speed were sensed by the simulated LRR and input into the automated longitudinal control for actuation of the powertrain and braking subsystems. The LRR is able to sense the relative speed (v_{radar}) between the automated vehicle and the target vehicle. For use in the automated longitudinal control, the actual speed of the target vehicle (v_{target}) could be determined by the expression in Equation (7).

$$v_{target} = v_x + v_{radar} \tag{7}$$

The visual feedback for detection of the target vehicle developed in dSPACE ControlDesk is shown in Figure 11.

The headway maintenance functionality of the automated longitudinal control had three settings: far, medium, and near. The headway distance ($d_{headway}$) of the automated longitudinal control was a function of the target vehicle speed (v_{target}) and time to collision (t_{TTC}) as shown in Equation (8).

$$d_{headway} = v_{target} t_{TTC} \tag{8}$$

The headway settings and their corresponding times to collision used in determining the headway distance are shown in Table 3.

The headway settings for human driver used visual feedback developed in dSPACE ControlDesk are shown in Figure 12.

Using kinematic equations, the acceleration/deceleration request for the automated longitudinal control could be determined from the speed and distance detected by the LRR used in combination with the vehicle speed and headway distance as shown in Equation (9).

$$a_{x,req} = \frac{\left(v_{target}^2 - v_x^2\right)}{2\left(d_{target} - d_{headway}\right)} \tag{9}$$

TABLE 3 Headway settings for automated longitudinal control

Headway Setting	Time to Collision
Near	1 s
Medium	1.5 s
Far	2.1 s

FIGURE 11 Target vehicle detection for human driver visual feedback.

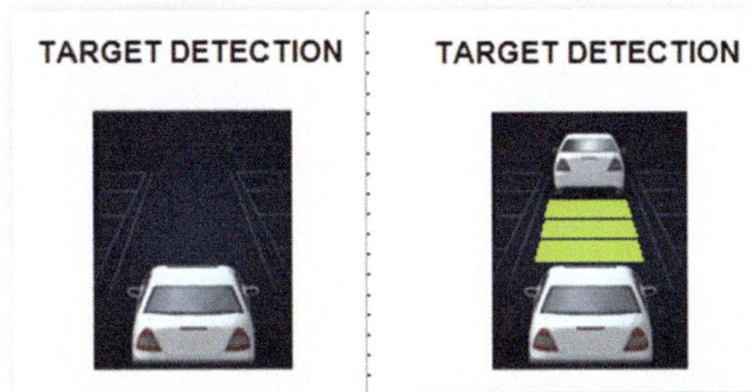

Automated Lateral Control

Steering is the primary function for lateral control. The steering subsystem of a vehicle is typically responsible for this function. The control objective of lateral automation systems is typically to keep the vehicle within a given lane such that the vehicle follows a reference trajectory or path [11].

Trajectory-tracking or path following is a common control problem regarding automated lateral control. Trajectory-tracking or path following typically involves the development and use of a control law that can provide control and actuation of steering such that the vehicle is able to trace a reference geometric path [11, 12]. The automated lateral control used in this research for path following was based on a sixth-order non-linear control-oriented model proposed by

Freund and Mayr for active front wheel steering [23]. The primary purpose of this control-oriented model was to achieve path following by varying the front wheel steering

FIGURE 12 Headway settings for human driver visual feedback.

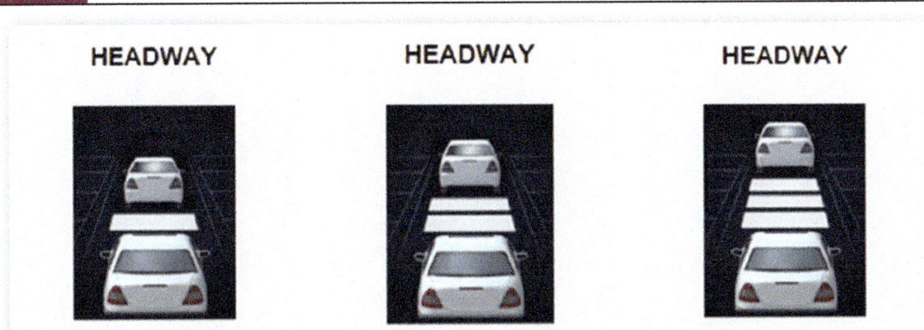

angles and drive/braking force [23]. The sixth-order non-linear control-oriented model based on the force balances with the front longitudinal and air resistance not considered is given by Equations (10), (11), (12), (13), (14), (15).

$$\dot{\beta} = -\dot{\omega}_Z + \frac{1}{m_v v_G}[-F_{lr}\sin(\beta) + S_v\cos(\delta_w - \beta) + S_h\cos(\beta)] \tag{10}$$

$$\dot{\psi} = \dot{\psi} \tag{11}$$

$$\ddot{\psi} = \frac{1}{I_{zz}}\left[l_f S_v cos(\delta_w) - l_r S_h\right] \tag{12}$$

$$\dot{v}_G = \frac{1}{m_v}\left[F_{lr}cos(\beta) - S_v sin(\delta_w - \beta) + S_h sin(\beta)\right] \tag{13}$$

$$\dot{X} = v_G cos(\psi + \beta) \tag{14}$$

$$\dot{Y} = v_G sin(\psi + \beta) \tag{15}$$

The states of sideslip angle (β), yaw angle (ψ), yaw rate ($\dot{\psi}$) and velocity (v_G) in combination with the vehicle trajectory at the center of gravity (X, Y) form the basis of the control-oriented model. The equations are also expressed in terms of the steering angle (δ_w) due to lateral vehicle dynamics. Rear longitudinal force (F_{lR}) and front lateral force (S_v) are considered as the control inputs of the system which are needed for developing a path following control law. Moreover, vehicle yaw moment of inertia (I_{zz}), total vehicle mass (m_v), and distances between center of gravity and front/rear axles (l_f, l_r) represent the constant vehicle geometry and body parameters used in this model. The expressions for the front lateral force (S_v) and rear lateral forces (S_h) in terms of the cornering stiffnesses ($C_{\alpha f}$, $C_{\alpha r}$) and states are shown in Equations (16)-(17).

$$S_v = C_{\alpha f}\left(\delta_w - \beta - \frac{l_f\dot{\psi}}{v_G}\right) \tag{16}$$

$$S_h = C_{\alpha r}\left(-\beta + \frac{l_r\dot{\psi}}{v_G}\right) \tag{17}$$

Relatively small steering angle (δ_w) and sideslip angle (β) were assumed for simplification of the non-linear state space representation of the model, which is shown in Equation (18).

$$\dot{x}(t) = \begin{bmatrix} -x_3 + \frac{S_h}{m_v x_4} \\ x_3 \\ -\frac{S_h l_r}{I_{ZZ}} \\ 0 \\ x_4 cos(x_1 + x_2) \\ x_4 sin(x_1 + x_2) \end{bmatrix} + \begin{bmatrix} \frac{1}{m_v x_4} & -\frac{x_1}{m_v x_4} \\ 0 & 0 \\ \frac{l_f}{I_{zz}} & 0 \\ 0 & \frac{1}{m_v} \\ 0 & 0 \\ 0 & 0 \end{bmatrix} u(t) \tag{18}$$

Where:
$x(t) = \left[\beta(t)\ \psi(t)\ \dot{\psi}(t)\ v_G(t)\ X(t)\ Y(t)\right]^T$
$y(t) = [X\ Y]^T$
$u(t) = [S_v(t)\ F_{lR}(t)]^T$

To understand the effect of the control inputs, rear longitudinal force (F_{lR}) and front lateral force (S_v), on the vehicle accelerations, the relative degrees of the system outputs in terms of the vehicle trajectory (X, Y) were analyzed. The relative degree of a system is defined as the number of times the outputs must be differentiated before the inputs appear explicitly. Since the relative degree of the system is two, the control inputs affect the vehicle accelerations directly as shown in Equations (19)-(20).

$$\ddot{y}_1 = -\frac{1}{m_v} S_{12} S_h - \frac{1}{m_v} S_{12} u_1 + \frac{1}{m_v}\left(C_{12} + x_1 S_{12}\right) u_2 \tag{19}$$

$$\ddot{y}_2 = -\frac{1}{m_v} C_{12} S_h + \frac{1}{m_v} C_{12} u_1 + \frac{1}{m_v}\left(S_{12} - x_1 C_{12}\right) u_2 \tag{20}$$

A feedback linearization control law of the form, $\ddot{y} = q(x) + S(x)u$, in terms of the vehicle longitudinal and lateral accelerations at the center of gravity, was proposed by Freund and Mayr to enable path following, where the expressions for $q(x)$ and $S(x)$ are shown in Equations (21)-(22) [23].

$$S(x) = \frac{1}{m_v}\begin{bmatrix} -S_{12} & C_{12} + x_1 S_{12} \\ \cos(x_1 + x_2) & S_{12} - x_1 C_{12} \end{bmatrix} \tag{21}$$

$$q(x) = \frac{1}{m_v}\begin{bmatrix} -S_{12} \\ C_{12} \end{bmatrix} S_h \tag{22}$$

Where:
$S_{12} = \sin(x_1 + x_2)$
$C_{12} = \cos(x_1 + x_2)$

The tracking error and tracking error dynamics between the actual trajectory (X, Y) and the desired trajectory (X_d, Y_d) of the vehicle were critical in determining the expressions for the control inputs (u). The expressions for the tracking error and tracking error dynamics are shown in Equations (23)-(24).

$$e = y_d - y = \left[X_d - X\ \ Y_d - Y\right]^T \tag{23}$$

$$\ddot{e} + \alpha\dot{e} + \lambda e = 0 \tag{24}$$

α and λ are parameters in the control law and the control law inputs (u) that can be tuned such that the error dynamics for the system are sufficiently fast and stable. The expressions for the control inputs (u) are given in Equations (25)-(26).

$$u_1 = -S_h + m_v\left[\left(x_1 C_{12} - S_{12}\right)\bar{y}_1 + \left(C_{12} + x_1 S_{12}\right)\bar{y}_2\right] \tag{25}$$

$$u_2 = m_v\left[C_{12}\bar{y}_1 + S_{12}\bar{y}_2\right] \tag{26}$$

Where:
$\bar{y}_1 = \lambda\omega_1 - \alpha(x_4 C_{12}) - \lambda x_5$
$\bar{y}_2 = \lambda\omega_2 - \alpha(x_4 S_{12}) - \lambda x_6$

FIGURE 13 Simulink-CarSim interface for automated lateral control [17].

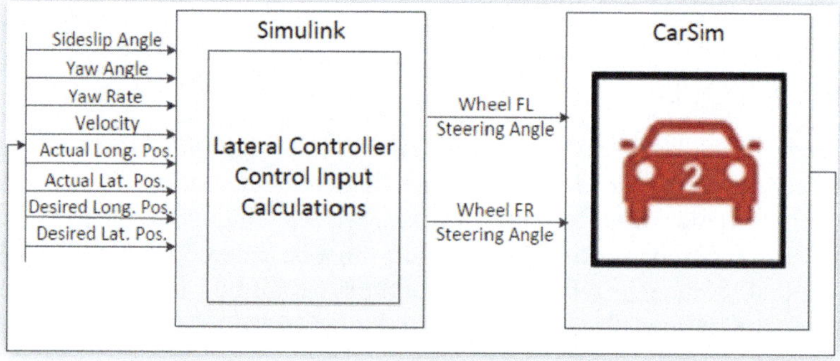

$$S_h = C_{\alpha r}\left(-x_1 + \frac{l_r x_3}{x_4}\right)$$

$$\omega = y_d + \begin{bmatrix} 1/\lambda & 0 \\ 0 & 1/\lambda \end{bmatrix}\left(\begin{bmatrix} \alpha_{11} & 0 \\ 0 & \alpha_{12} \end{bmatrix}\dot{y}_d + \ddot{y}_d\right)$$

The expression for the front steering wheel angles can be computed using the control input expressions as shown in Equation (27).

$$\delta_w = \frac{u_1}{C_{\alpha f}} + x_1 + \frac{l_f x_3}{x_4} \tag{27}$$

As the scope of this model was to enable path following control, only the controller part of the model proposed by Freund and Mayr was used and implemented [23]. Correspondingly, due to accuracy reasons, the six states of the model $\left(\left[\beta \;\; \psi \;\; \dot{\psi} \;\; v_G \;\; X \;\; Y\right]^T\right)$ and the desired longitudinal and lateral trajectories of the path (X_d, Y_d) were not used from the estimation portion of the control-oriented model, and as a result, were used directly from the CarSim plant model. The calculations pertaining to the control law of the automated lateral control were conducted in Simulink. The front steering wheel angles were passed through the CarSim-Simulink interface. As active rear wheel steering was outside the scope of this paper, the rear steering wheel angles were ignored in the implementation of the automated lateral control. The CarSim- Simulink interface used for the real-time implementation of the automated lateral control is shown in Figure 13.

HIL SAE Level 2 Automation

Decision and control form a key component of the architecture of a vehicle automation system. Decision - making involves the automation system deciding upon on the appropriate actions to take based on information from various sensors and sources of the surrounding environment [11]. The automated control system via decision-making is also responsible for generating control commands to the actuators. In the case of longitudinal automation, this command is a desired acceleration and deceleration sent to the powertrain and brakes subsystems respectively. For lateral automation, the control command to the steering actuator is a desired front wheel steering angle sent to the steering subsystem.

In order to attain SAE Level 2 automation, dependability of the system is as important as decision and control. The dependability of a system can be expressed in terms of fault prevention and fault tolerance. The former concept is the prevention of the occurrence or introduction of faults due to development and design processes [11]. The purpose of the latter concept, fault tolerance, is to avoid failures when faults are present at the component, subsystem or system level in order to ensure safety [11]. The strictness of dependability requirements is proportional to level of automation. A method of attaining the required level of fault tolerance is the inclusion of alternative control systems or features that can ensure safety of the system in case the nominal automation system fails [24]. This alternative control system can include features for fallback performance of system such that the drivability capability of the vehicle is still maintained although the performance may be degraded [25].

Two common attributes closely related to dependability are availability, which involves readiness for correct service, and reliability, which involves continuity of correct service [11]. The improvement or degradation of one of these attributes may not be proportional to the other attribute. In certain cases, it may be better to improve reliability of the system at the expense of availability. With respect to automation, this may involve disabling or degrading automation under some conditions instead of continued operation with poor reliability to ensure safety. In these types of situations, the driver can take over complete or partial control of the vehicle when automation is disabled [26, 27].

A key issue that has gained attention regarding SAE Level 2 autonomous vehicles is that of active driver re-engagement in a takeover scenario to assume active control of the vehicle [27]. These types of scenarios are considered to be highly complex and depend on the human driver's ability to re-engage with the system in order to take over driving capability due to the occurrence of faults within the automated controls system [8].

In a vehicle, the faults associated with the automated longitudinal control may be initiated by either the associated actuators or some failure in the physical subsystem for powertrain or brakes [14]. In a vehicle, the faults associated with the automated lateral control, however, may be typically due to failures of the sensors used to determine the spacing between vehicles and tracking the reference path. [14].

This research is primarily concerned with the HIL implementation of an SAE Level 2 autonomous vehicle to enable robust control of an automated vehicle system subject to faults to either the longitudinal control or lateral control. Due to the limited scope of this research, only faults which still allow continued operation of the vehicle will be considered and faults that involved complete failure of subsystems are not considered [14].

SAE Level 2 autonomous vehicles utilize shared authority when the human driver cedes active primary control in certain limited driving situations such that the human driver is disengaged from physically operating the vehicle [28]. Due to the nature of an SAE Level 2 autonomous vehicle, there will always be situations in which the automation will not be able to handle system limits and result in disengagements. This is handled by the concept of shared authority or control which involves transitioning between the various levels of automation [28]. Shared authority accomplished these transitions such that all tasks required for proper operation of the vehicle are properly distributed between the human driver and the automated system. This interaction between the human driver and the automated system for an SAE Level 2 autonomous vehicle is accomplished by a feedback mechanism as shown in Figure 14.

Most automation systems have automated controls in which all of the control authority for both lateral and longitudinal control is handed back to the human driver when a fault is detected. This research proposes adding an extra layer of fallback performance that includes intermediate states of SAE Level 1 automation for automated longitudinal control and automated lateral control [28]. The first of these intermediate states involve using automated lateral control only as a fallback to a fault in the automated longitudinal control when the vehicle is in fully autonomous mode. The second of these

FIGURE 14 Human driver - automated system interaction for SAE Level 2 autonomous vehicle [28].

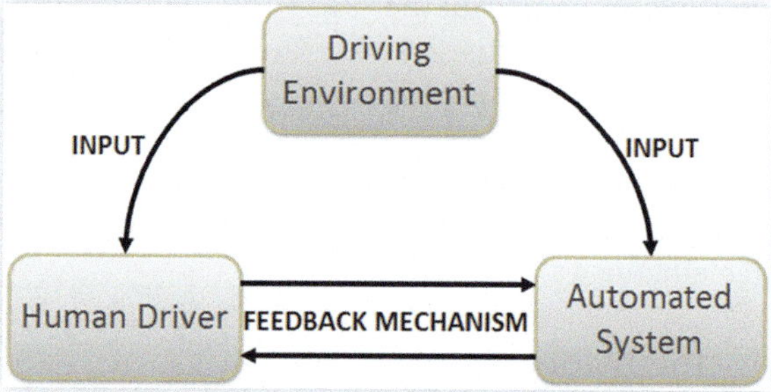

intermediate states involve using automated longitudinal control only as a fallback to a fault in the automated lateral control when the vehicle is in fully autonomous mode. These intermediate states can provide either lateral or longitudinal control and thereby stability while the human driver is re-engaging to take over controls. Another benefit of this system would be the independence of the lateral and longitudinal controls such that loss of one will not affect the other, while also maximizing the potential benefits each subsystem has regarding either lateral or longitudinal automation. The transition of states for the automated control showcasing the fallback states is shown in Figure 15.

The automated system could be engaged from manual driving mode by pressing the steering wheel button and could voluntarily be disengaged by either depressing the accelerator pedal or brake pedal. These actions indicated human driver intention to take over complete vehicle control. In case of a fault in the automated longitudinal control, the automated system would transition into SAE Level 1 operation by only allowing the automated lateral control, while the human driver would take over vehicle control of acceleration and braking. However, in the case where there was a fault in the automated lateral control, the automated system would transition into SAE Level 1 operation by only allowing the automated longitudinal control, while the human driver would take over vehicle control of steering. The automated system could also transition into this fallback state when the human driver applied a steering wheel torque indicating intention to take over steering only. The human driver could force the fallback states to transition to the disengaged state by depressing the brake pedal. By depressing the accelerator pedal from the automated longitudinal control only fallback state, the autonomous mode could also be disengaged. Conversely, if the faults in the subsystems dissipated, the autonomous mode could be engaged directly from the fallback states by pressing the steering wheel button. However, to transition from the automated longitudinal control only mode to autonomous mode, the steering wheel must also be centered by the human driver before engagement.

Another important issue regarding SAE Level 2 autonomous vehicles is that of driver situation awareness. Due to increase in automation of vehicle, a human driver's situation awareness of the surrounding environment may decrease, potentially adversely affecting

FIGURE 15 Transition of states for automated control of SAE Level 2 autonomous vehicle.

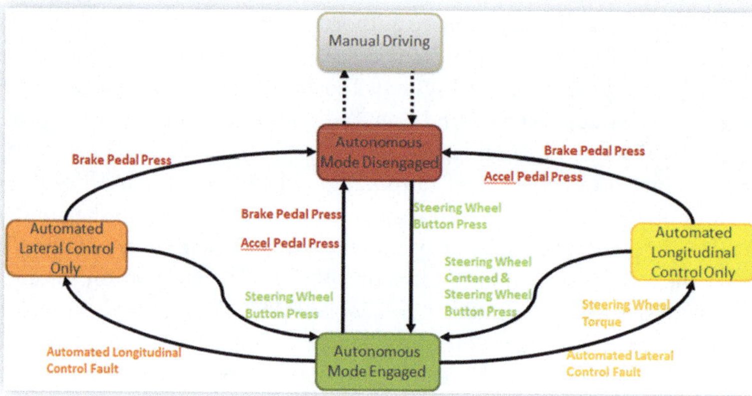

FIGURE 16 Human driver-automated system shared authority for SAE Level 2 autonomous vehicle [28].

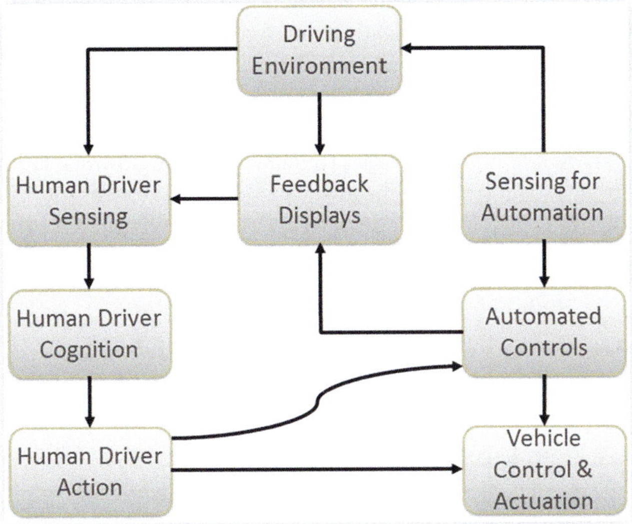

the response during re-engagement due to subsystem or system faults. According to the NHTSA a warning should be provided to the driver in an autonomous vehicle prior to any transition or take-over [29]. A sufficient warning to take-over interval is essential to allow the driver to fully assess the situation and reengage in the driving task. In order to attract the human driver's attention in response to such events, there is a critical need for an effective feedback of the automation functionality to the human driver [30, 31]. This feedback allows for proper functionality of the shared authority between the human driver and automated system the proper vehicle actuation and control is allowed as shown in Figure 16.

One type of effective feedback is visual feedback. Most information and visual feedback from the vehicle is transmitted to the human driver as part of the Instrument Panel Cluster (IPC). The simulated IPC developed in dSPACE ControlDesk is shown in Figure 17.

However, for automated controls, more detailed forms of visual feedback were required to be developed. For one form of visual feedback for the human driver, a series of simulated Light-Emitting Diodes (LEDs) which could be cycled between different colors and on/off states were used to indicate the transition states. For the autonomous mode engaged state, all four LEDs lit up to the green color indicating the successful engagement. All four LEDs lit up to the yellow and orange colors for the automated

FIGURE 17 Simulated instrument panel cluster (IPC).

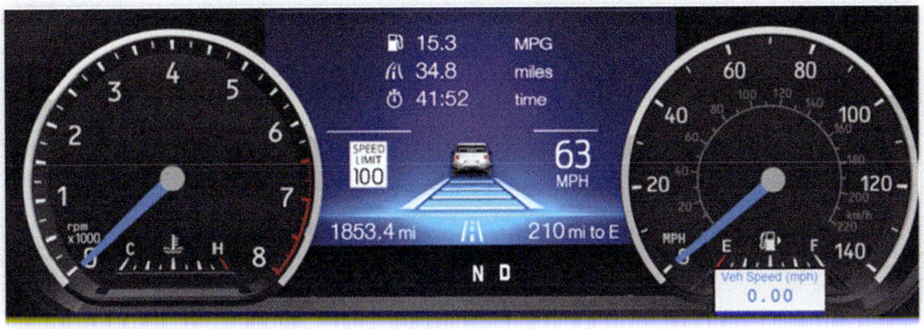

FIGURE 18 Visual feedback for the states of automated control for SAE Level 2 autonomous vehicle.

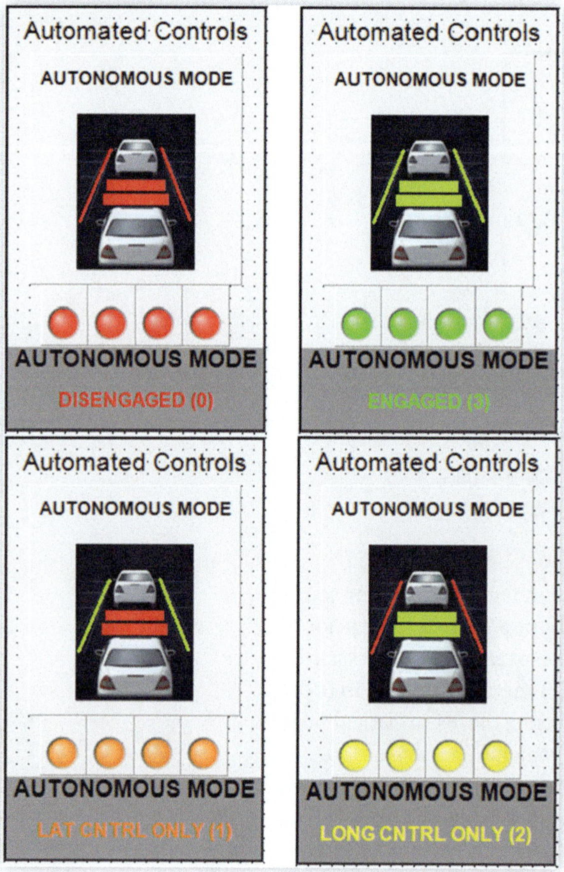

longitudinal control and automated lateral control available only modes respectively. For the autonomous mode disengaged state, all four LEDs lit up to the red color indicating the urgency of the situation.

Another form of visual feedback included a human machine interface showing the engagement or disengagement of the automated lateral and longitudinal controls. The automated longitudinal control was shown by a headway gap visual representation. The automated lateral control was shown by the visual representation of two lanes. Both forms of visual feedback for the different states developed in dSPACE ControlDesk are shown in Figure 18.

Audible warnings are another type of discrete feedback of automation functionality [26]. It has been shown that audible warnings can aid improving the human driver response to safety critical events. It has also been shown that audible warnings tend to improve the human driver response time as compared to systems in which there are no warnings provided [32, 33, 34].

For SAE Level 2 autonomous vehicles, in situations where the human driver is engaged in non-driving tasks, the effects of audible warnings drastically improve drivers' reaction time such that they draw the human driver's attention back into control faster than use of only simplistic visual feedback of the warning [26]. In this research the purpose of addition of audible warnings was to alert driver to a change in the state of the autonomous vehicle system. Audible warnings and voice commands corresponding to the different states were added. For the states, audible warnings were helpful in providing the human driver with feedback so that the human driver is alerted to re-engaging into the vehicle control loop such that the vehicle control can be restored to drive the vehicle following a fault in one of the automated controls. For the autonomous mode engaged state, a single tone sound was used in conjunction with a voice command stating 'Autonomous Mode Engaged' to provide audible feedback. For the fallback state transitions, a two tone sound was used in conjunction with voice commands stating 'Longitudinal Control Available Only' and 'Longitudinal Control Available Only' to provide audible feedback. For the autonomous mode disengaged state, a four tone rapid sound was used in conjunction with a voice command stating 'Autonomous Mode Disengaged' to provide audible feedback and indicate the urgency of the situation.

Simulations and Test Results

The simulations were conducted on the Simulink-CarSim HIL platform representation of 2017 Ford Fusion Hybrid. The actual vehicle and its virtual vehicle representation are shown in Figure 19.

The vehicle body parameters for the 2017 Ford Fusion used in the Simulink-CarSim HIL platform are shown in Table 4.

The goals of the simulations and testing were to validate the automated lateral control and automated longitudinal control components in real-time on the HIL implementation of an SAE Level 2 autonomous vehicle. This allowed for the use of an accessible, safe and inexpensive simulation platform for controls testing of automated controls systems. To validate the automated longitudinal control, simulations were conducted on a flat straight-line track. The automated lateral control was validated on a test track which consisted of turns and straightaway sections. The virtual test tracks used for the simulations are shown in Figure 20.

Moreover, simulations were also conducted in which an SAE Level 2 automated control system was subjected to lateral and longitudinal subsystem faults to understand the fallback capability of the intermediate states of the designed automated control system. This analysis showcased the importance of the lateral and longitudinal controls being independent and thereby increasing reliability in case of failure of one subsystem. The responses of the intermediate states were compared by showcasing continued operation of the subsystems.

The validation of the automated lateral and longitudinal controls was accomplished with the use of the Root Mean Square Error (*RMSE*) and the coefficient of determination. The Root Mean Square Error (*RMSE*) is defined as the sampled standard deviation between varied dataset and a nominal dataset. The coefficient of determination is a metric of the quality of a data fit with respect to a nominal dataset. A value close to 1 corresponds to an increased quality of the data fit and decreased variation between two datasets. The expressions for the Root Mean Square Error (*RMSE*) and the coefficient of determination (R^2) are shown in Equations (28)-(29).

$$RMSE = \sqrt{\frac{1}{n}\sum_{i=1}^{n}(V_i - N_i)^2} \tag{28}$$

$$R^2 = 1 - \frac{\sum_{i=1}^{n}(V_i - N_i)^2}{\sum_{i=1}^{n}\left(V_i - \frac{1}{n}\sum_{i=1}^{n}V_i\right)^2} \tag{29}$$

Where:
N_i = Nominal value
V_i = Varied value
n = Total number of data points

FIGURE 19 2017 Ford Fusion and corresponding CarSim representation [17].

TABLE 4 Vehicle body parameters of 2017 Ford Fusion [10]

Parameter	Value
m_v	1815 kg
l_F	1.1164 m
l_R	1.7039 m
I_{ZZ}	2747.3 kgm^2
$C_{\alpha f}$	61058 N/rad
$C_{\alpha r}$	41378.5 N/rad

FIGURE 20 Virtual test tracks in CarSim.

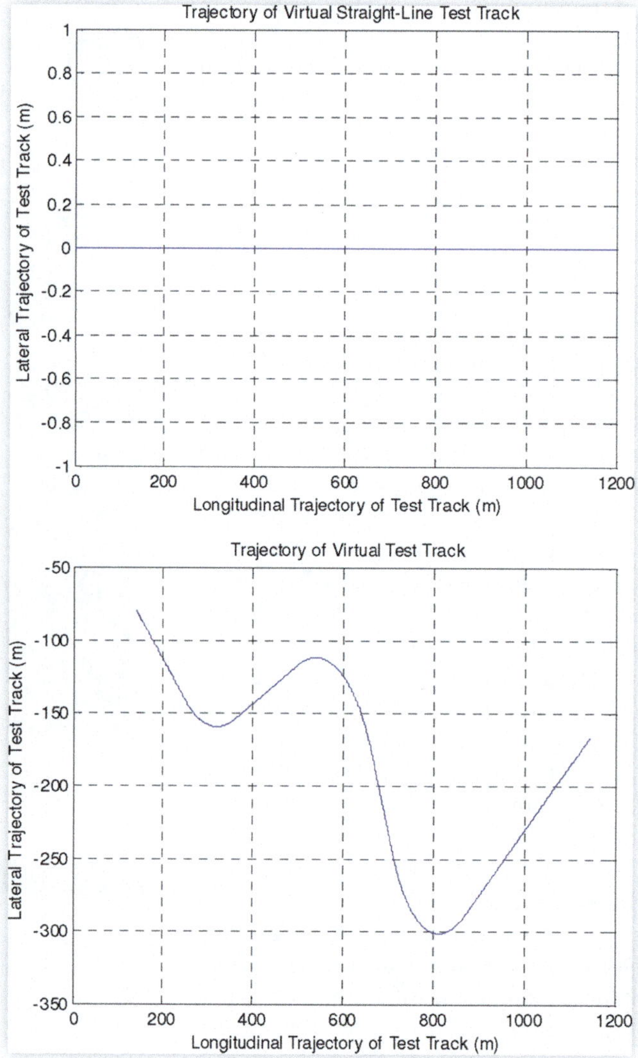

The simulations and test results for the automated longitudinal control, automated lateral control, and fault tolerance of the transition states of an SAE Level 2 autonomous vehicle are discussed below.

Automated Longitudinal Control Simulations & Test Results

For automated longitudinal control validation, the HIL implementation of an SAE Level 2 autonomous vehicle was subjected to three speed ranges on the straight-line virtual track. The three speed ranges were 40 km/hr, 80 km/hr, and 120 km/hr indicative of low, medium, and high speeds. Maintaining a desired speed and keeping a desired distance to other vehicles were the two key components for automaton of longitudinal control. The test results for maintaining a desired speed are shown in Figure 21 and Table 5. The test results for maintaining a desired distance to the target vehicle are shown in Figures 22, 23, and Table 6.

From the plots for maintaining the desired speed over the three speeds ranges, it was shown that that automated longitudinal control allowed the actual vehicle speed to track the reference set-point speed after accelerating the vehicle to match the reference set-point speed. At steady-state, there was relatively small error between the vehicle speed response and the reference set-point speeds. This was showcased by the relatively small Root Mean Square Error values of 0.0987, 0.8742 and 1.1592 across the three speed ranges respectively. Moreover, at steady-state the coefficient of determination values of 0.9996, 0.9999, and 0.9999 across the three speed ranges respectively further corroborated the observation of regarding the performance of the automated longitudinal control for maintaining the reference set-point speed.

The plots regarding the maintenance of a desired headway distance showcased the ability of the automated longitudinal control to decelerate the vehicle once the target vehicle was detected by the LRR such that it could match the speed of the target vehicle with minor variations by maintaining an appropriate headway distance according to the selected headway settings. In the simulations across the three speed ranges, the target vehicle was driven at 10 km/hr such that the required headway distance according to the far headway setting was 6.3 m based on the time to collision specification of 2.1 s. The automated vehicle was able to decelerate from all three speed ranges to match the target vehicle speed of 10 km/ hr and was able to maintain the requisite headway distance of 6.3 m between the automated vehicle and the target vehicle with minor variations. The values of the Root Mean Square Error of 0.3239, 0.2083, and 0.1752 for the headway distance response verified this observation at steady-state. Moreover, coefficient of determination values of 0.9995, 0.9998, and 0.9998 for the headway distance response at steady-state indicated the high accuracy of the data fit between the actual distance to the target vehicle and the required headway distance.

The high quality of data fit between the response and the reference speeds in unison with the distance response and set headway validated both components of the automated longitudinal control. The relatively small errors observed may be due to the nature of the automated longitudinal control being a reactive control instead of predictive control. The automated longitudinal control reacts to the varying conditions and adjusts the response accordingly, therefore, leading to minor delays and latencies in the system response and a lack of perfect tracking.

FIGURE 21 Vehicle speed test results of the automated longitudinal control.

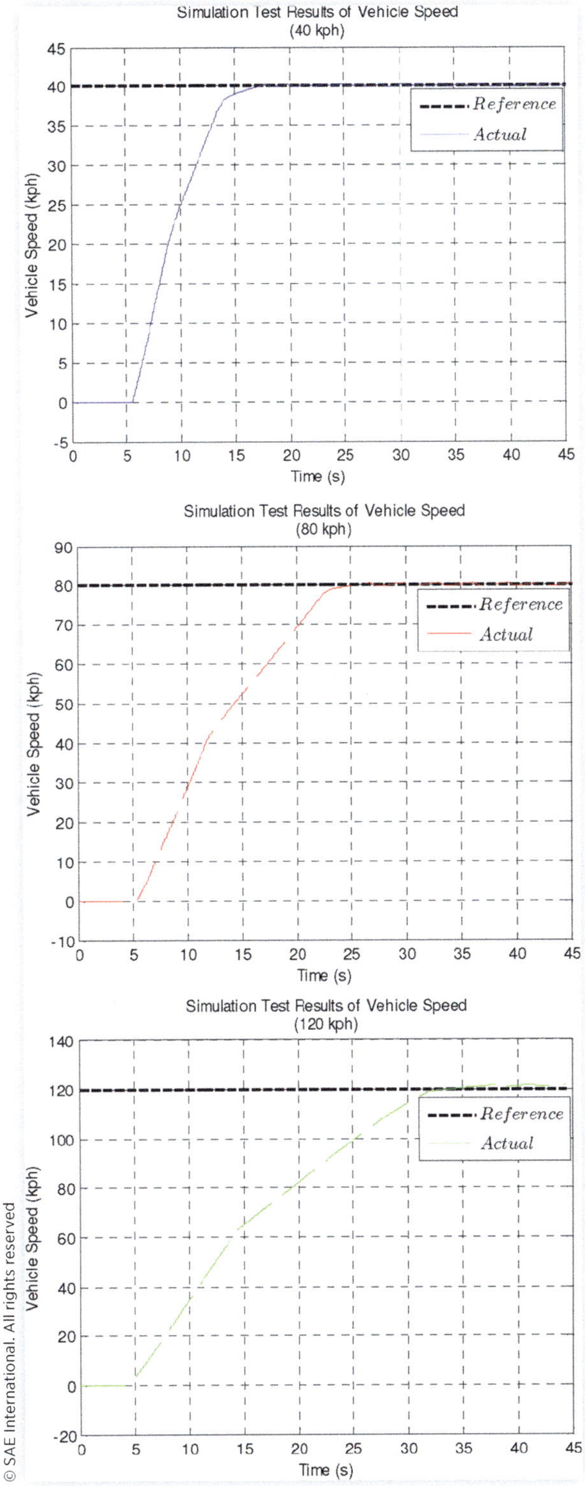

CHAPTER 8

TABLE 5 Automated longitudinal control test results

Speed Range	Root Mean Square Error	Coefficient of Determination
40 km/hr	0.0967	0.9996
80 km/hr	0.8742	0.9999
120 km/hr	1.1592	0.9999

Moreover simulations were also used to showcase the different headway settings available to the human driver while during operation of the automated longitudinal control. In the simulations, the target vehicle speed was set to 10 km/hr such that the headway settings were alternated between the near, medium and far settings. The required headway distance was alternated between 3 m, 4.5 m, and 6.3 m based on the times to collision specifications of 1 s, 1.5 s, and 2.1 s respectively. The simulations of the headway settings are shown in Figure 24.

Automated Lateral Control Simulations & Test Results

The automated lateral control was validated by validation of the path following trajectory of SAE Level 2 autonomous vehicle on the virtual test track which consisted of turns and straightaway sections. Simulations were executed over three speed ranges. The three speed ranges were 40 km/hr, 60 km/hr, and 80 km/hr indicative of low, medium, and high speeds. Higher speed ranges such as those used for testing the autonomous longitudinal control could not be tested due to vehicle instability during the turning sections of the virtual test track. For correlation, the steering wheel angle and lateral acceleration responses were compared. The test results for the automated lateral control over the three speed ranges are shown in Figure 25 and Table 7.

For analysis of the automated lateral control, the Root Mean Square Error values and the coefficient of determination values were determined by comparing desired vehicle trajectory of the virtual test track with the actual trajectory traced by the automated vehicle. The above plots show that the automated lateral control can track the reference trajectory of the virtual test track with relatively small error. This was verified by the small Root Mean Square Error values of 1.0136, 1.0328, and 1.0359 for the longitudinal trajectory in combination with the values of 0.5131, 0.5231, and 0.5248 for the lateral trajectory across the three speed ranges. In addition, coefficient of determination values of 0.9999 for the longitudinal and lateral trajectories across the three speed ranges respectively indicated high accuracy of the data fit between the reference path trajectory and the automated vehicle trajectory. The source of the small variations noticed may be due to the tire dynamics utilized in the simulations. Due to the tendency of tires to exhibit nonlinearities at higher speeds, the nonlinear nature of the tire lateral and longitudinal force curves may not be taken into account for the cornering stiffness values used in the simulations. The minor discrepancies observed may be indicative of a non-linear relationship between the longitudinal acceleration and steering angle which is not considered in the simulations. However, the high accuracy of data fit between the trajectory response and the reference trajectory validated the automated lateral control. The real-time performance of the automated lateral control discussed in this research has been described and accomplished in detail in [17].

FIGURE 22 Headway test results of the automated longitudinal control.

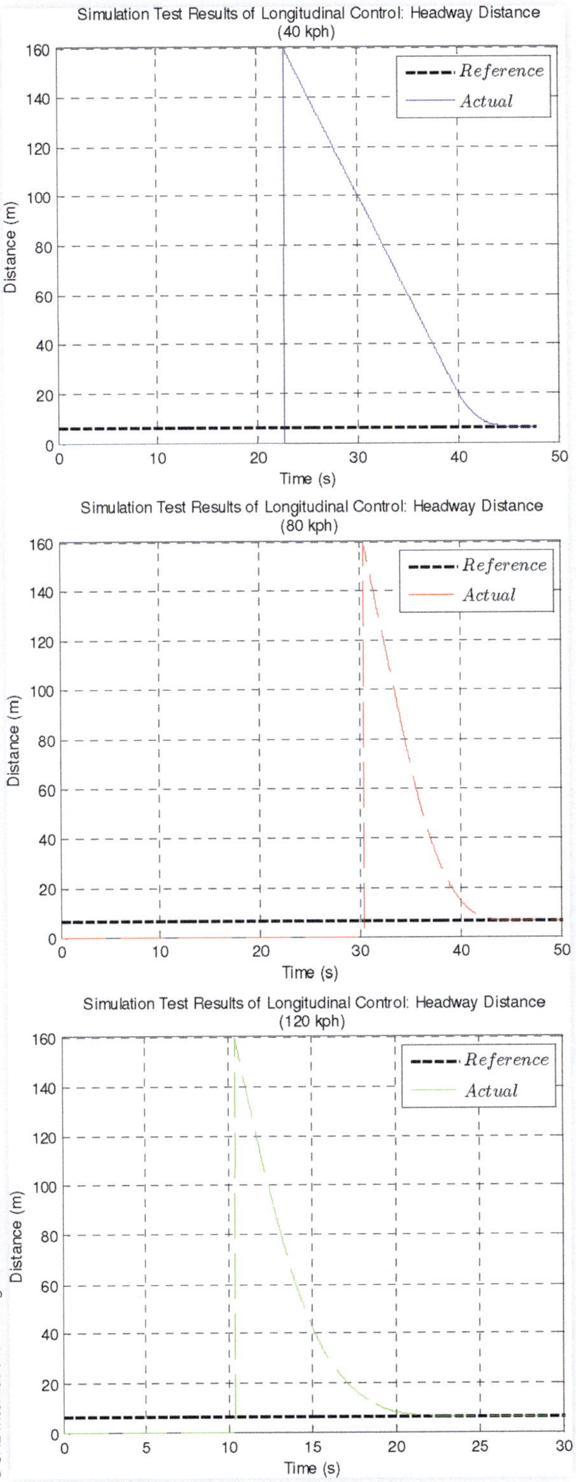

FIGURE 23 Vehicle speed test results of the automated vehicle and target vehicle.

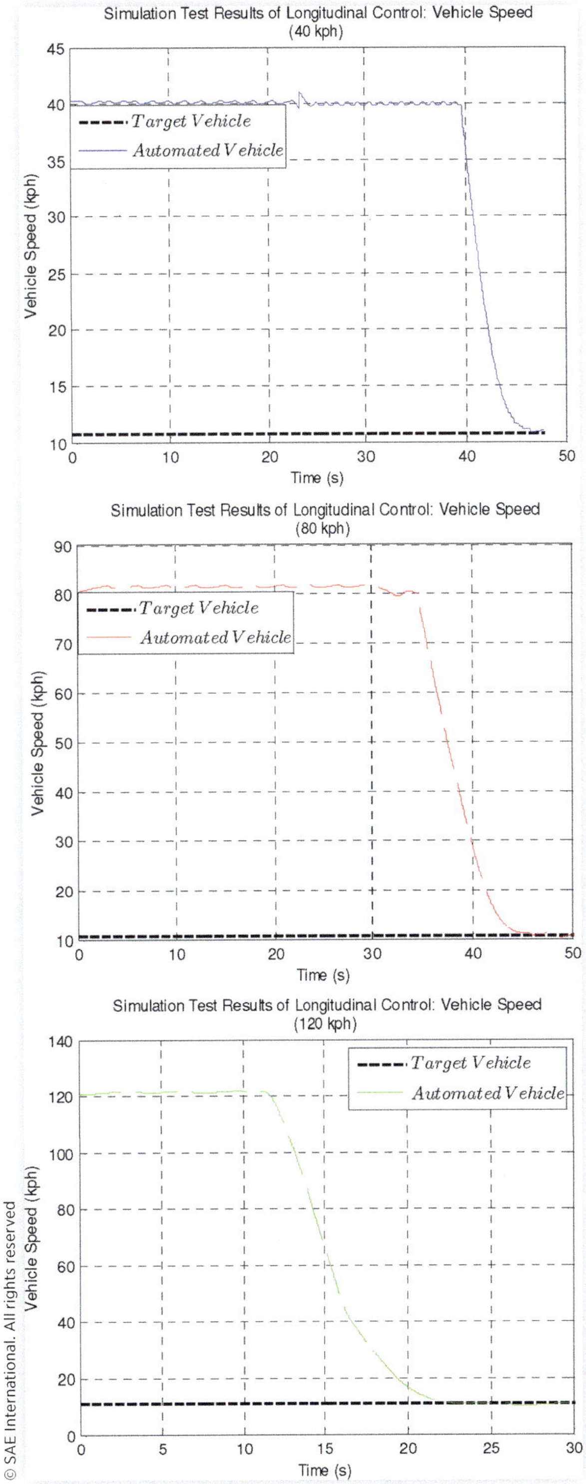

TABLE 6 Automated longitudinal control test results

Speed Range	Root Mean Square Error	Coefficient of Determination
40 km/hr	0.3239	0.9995
80 km/hr	0.2083	0.9998
120 km/hr	0.1752	0.9998

FIGURE 24 Different headway settings simulations.

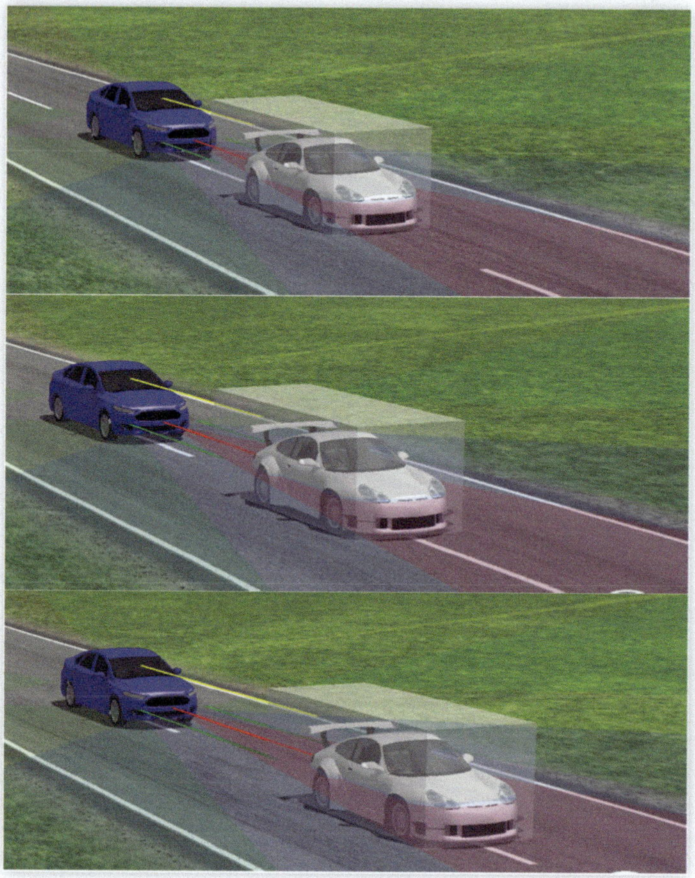

Subsystem Faults and Fallback Performance Simulation & Test Results

The fallback performance was evaluated and validated by introducing subsystem faults when the automated vehicle was in autonomous mode. This would force the automated system into the fallback states. The automated lateral control only fallback mode would be triggered in the event of a fault in the automated longitudinal control. The automated longitudinal control fallback mode only fallback mode would be activated due to the presence of a fault in the automated lateral control. The simulations were conducted at vehicle speeds of 40 km/hr and 80 km/hr on the straight-line virtual test track. The engagement of the autonomous mode is indicated by a green line on the plots whereas the disengagement of the autonomous mode and following transition into the fallback states is indicated by a red line on the plots. The test results for the fallback performance over the given speeds are shown in Figures 26, 27, 28, 29, Tables 8 and 9.

From the results, it can be seen that the automated lateral control continues to operate by enabling trajectory tracking even though there was a fault in the automated longitudinal control causing loss of the latter subsystem. During this scenario, automated system went to its fallback state at t = 30 s and t = 35.74 s for the 40 km/hr and 80 km/ hr vehicle speed cases respectively. As indicated by the plots, after the activation of the fallback state, the vehicle speed dropped due to disengagement of the automated longitudinal control. However the automated lateral control was observed to operate continuously in order to perform path following after the state transition to the fallback state. This was corroborated by Root Mean Square Error values of 0.0011 and 0.0007 for the vehicle response when being driven at 40 km/hr and 80 km/hr respectively. Moreover, corresponding coefficient of determination values of 0.9999 for these two cases further validated the automated lateral control response in the fallback state.

FIGURE 25 Path following test results of the automated lateral control.

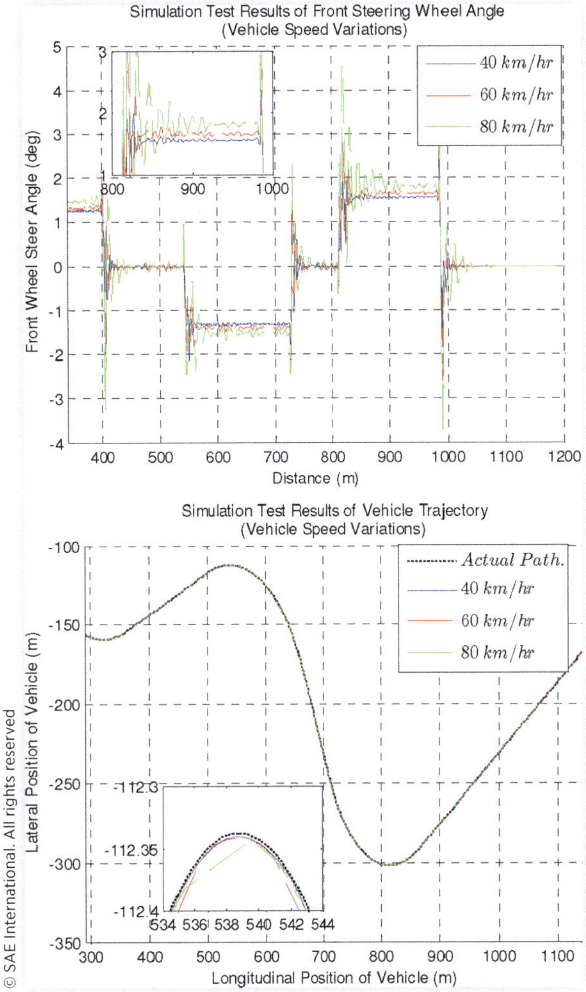

As observed from the above simulations and test results, the automated longitudinal control continued to function by maintaining the desired speed during the course of a fault being present in the automated lateral control causing transition to a SAE Level 1 automated state due to loss of the latter subsystem. The automated system was triggered to fall back to the automated longitudinal control available only mode. This transition can be observed at t = 22 s and t = 27.27 s for the two speed ranges respectively. As indicated by the plots, after the activation of the fallback state, the lateral trajectory of the vehicle fluctuated due to the re-engagement of the human driver to take over steering controls. However the automated longitudinal control was seen to function properly in order to maintain the desired speed function when the fallback state was triggered. Root Mean Square Error values of 0.1682 and 0.2799 for the vehicle response further verified this

TABLE 7 Automated lateral control test results

Speed Range	Root Mean Square Error		Coefficient of Determination	
	Long. Traj.	Lat. Traj.	Long. Traj.	Lat. Traj.
40 km/hr	1.0136	0.5131	0.9999	0.9999
60 km/hr	1.0328	0.5231	0.9999	0.9999
80 km/hr	1.0359	0.5248	0.9999	0.9999

FIGURE 26 Simulation test results of autonomous mode to lateral control only mode transition at 40 km/hr.

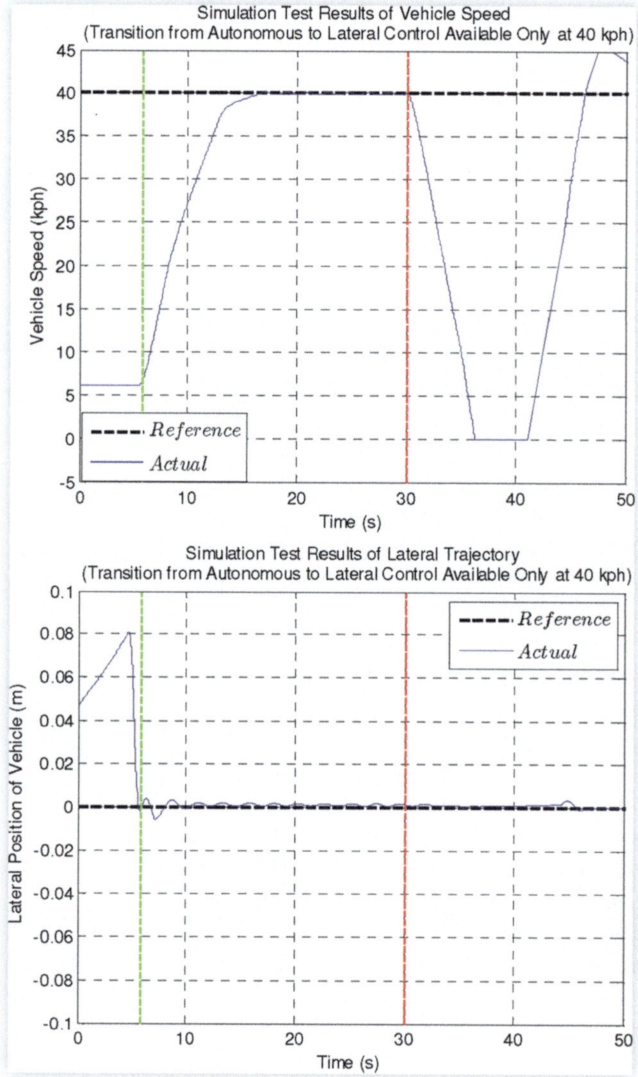

observation for the two speed ranges respectively. Additionally, the automated longitudinal control in this fallback state was validated using the corresponding coefficient of determination values of 0.9997 and 0.9995 for these two cases.

These simulations and test results showed the effects of simulated subsystem faults on the automated system. Moreover, this showcased the importance of repeatability of tests for design of experiments and robustness testing of controls. This also emphasized that testing of unsafe situations in which the automated system may transition to its fallback states in which either automated longitudinal control or automated lateral control is faulted can be done in a controlled and consistent simulated environment. Testing of these situations would otherwise be considered unsafe using in-vehicle testing due to the nature of faulting complete subsystem at high speeds. Additionally, this illustrated the benefits of a HIL simulation being more accessible and cost-effective for testing complex scenarios such as those described above.

FIGURE 27 Simulation test results of autonomous mode to lateral control only mode transition at 80 km/hr.

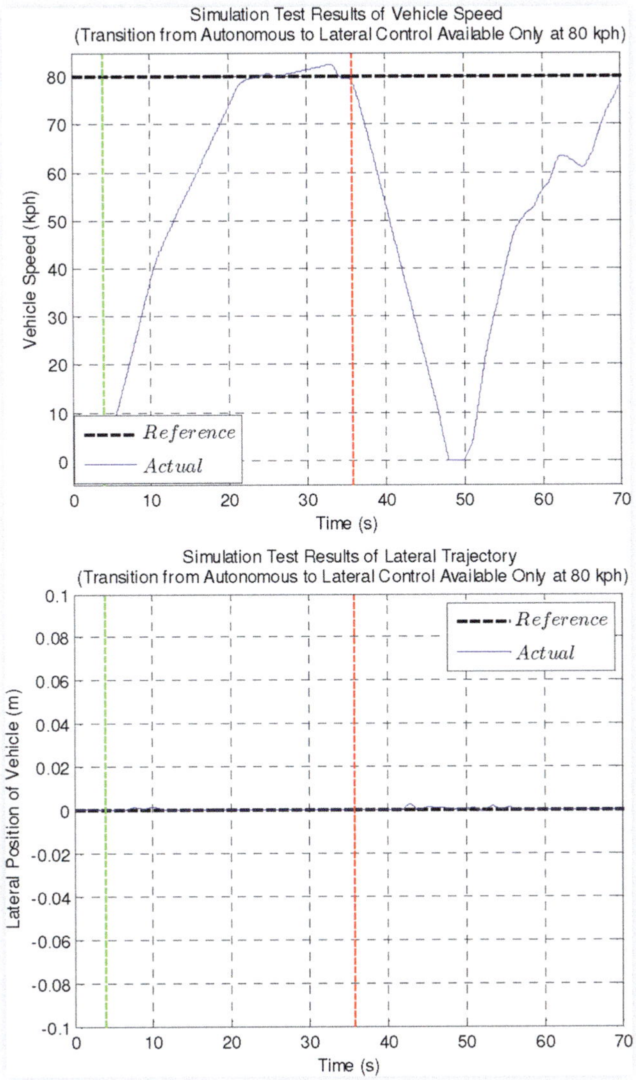

Conclusions

This paper illustrated the benefits of controls rapid prototyping and development using HIL simulation. This method of testing provided a consistent and controlled test environment in which robustness testing of controls for the human driver-automated system interaction could be tested in a cost-effective and accessible environment with a high measure of reliability. Studies have shown that HIL testing simulation methods can result in efficiency gains up to 50% in the controls development process with increased error reductions and increased acceleration in the maturity level of the controls [35]. Moreover, overall costs for validation of controls on using a HIL simulation using this approach are considerably lower as compared to in-vehicle testing. The cost comparison for a typical CarSim HIL setup and vehicle testing for controls validation is shown in Table 10 [17, 36].

FIGURE 28 Simulation test results of autonomous mode to longitudinal control only mode transition at 40 km/hr.

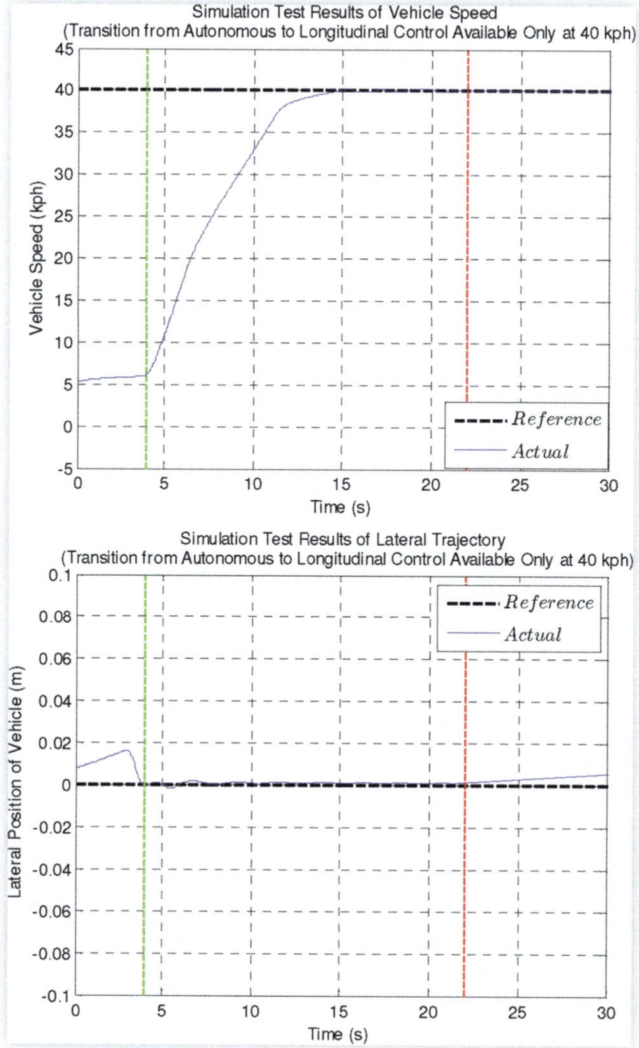

In this research, an SAE Level 2 automated system was developed and tested using a HIL setup. Moreover, the real-time performance of the longitudinal and lateral components of the automated controls was validated separately initially. This allowed for testing the performance of SAE Level 2 autonomous vehicle controls during takeover and human-driver re-engagement scenarios due to subsystem faults. Moreover this enabled the iterative development, testing and validation of the engagement of the fallback states in case of subsystem faults during autonomous operation. The SAE Level 2 autonomous vehicle HIL implementation provided a simulation platform for further analysis of vehicle automation and controls. A potential application of this setup could be utility as a driving simulator to understand and analyze the interaction between the automated system and the human driver for higher levels of automation. As SAE Level 2 and 3 autonomous vehicles rely on this interaction for proper operation, the HIL simulation could potentially be used to survey the time required by the human driver to take over when a takeover condition is initiated by the automated system.

FIGURE 29 Simulation test results of autonomous mode to longitudinal control only mode transition at 80 km/hr.

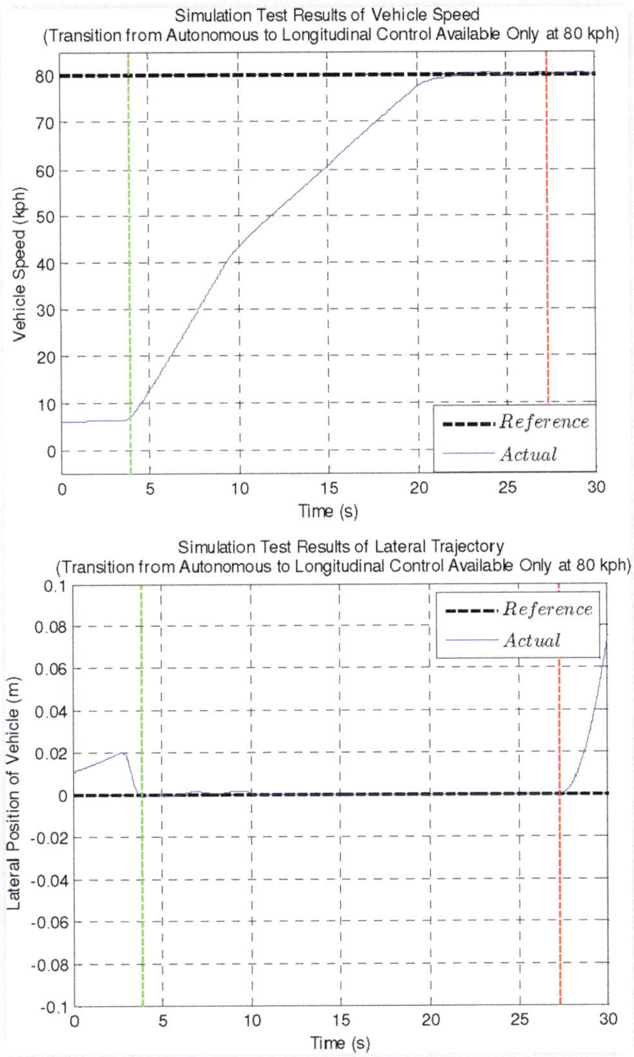

TABLE 8 Lateral control only available mode test results

Speed Range	Root Mean Square Error	Coefficient of Determination
40 km/hr	0.0011	0.9999
80 km/hr	0.0007	0.9999

TABLE 9 Longitudinal control only available mode test results

Speed Range	Root Mean Square Error	Coefficient of Determination
40 km/hr	0.1682	0.9997
80 km/hr	0.2799	0.9995

TABLE 10 Cost comparison for CarSim HIL simulation and vehicle testing [17, 36]

Vehicle Testing		CarSim HIL Simulation	
One-Time Costs		One-Time Costs	
Instrumentation	$50000	Lab construction	$120000
Recurring Costs		Recurring Costs	
Test vehicle and installation	$40000	Energy	$2000
Transportation	$20000	Installation	$5000
Fuel	$8000	Labor	$20000
Staff	$180000		

Future work could involve addition of production Adaptive Cruise Control (ACC) components such as the controls module and the radar for automated longitudinal control. Addition of camera-based lane-centering features along with a camera-in-the-loop hardware setup for the automated lateral control so that the accuracy of the HIL simulation is closer to that of the actual vehicle will be considered. Moreover, in order to progress towards higher levels of automation, AV target hardware and software for an SAE Level 4 autonomous vehicle which includes more complex software in terms of deep machine learning algorithms may be integrated with the current HIL platform to enable testing of complex scenarios which cannot be reproduced safely in the actual vehicle. This will allow further regression testing of automated controls at the system and subsystem level of the autonomous vehicle in a simulated environment [17].

Contact Information

Adit Joshi
Ford Motor Company
Dearborn, MI, USA
ajoshi23@ford.com

Abbreviations

ADAS - Advanced Driver Assistance Systems
AV - Autonomous Vehicle
HIL - Hardware-in-the-Loop
NHTSA - National Highway Traffic Safety Administration
ECU - Electronic Control Unit
PWM - Pulse Width Modulated
CAN - Controller Area Network
ECM - Engine Control Module
HCU - Hybrid Control Module
GSM - Gear Shift Module
TRCM - Transmission Range Control Module
GWM - Gateway Module
MABX - Microautobox
FHEV - Full Hybrid Electric Vehicle

LRR - Long-Range Radar
IPC - Instrument Panel Cluster
LED - Light-Emitting Diodes
RMSE - Root Mean Square Error

References

1. SAE International Recommended Practice, "Taxonomy and Definitions for Terms Related to Driving Automation Systems for On-Road Motor Vehicles," SAE Standard J3016, Rev. September 2016.
2. Singh, S., "Critical Reasons for Crashes Investigated in the National Motor Vehicle Crash Causation Survey," Traffic Safety Facts Crash Stats, National Highway Traffic Safety Administration, Report No. DOT HS 812 115, February 2015.
3. National Center for Statistics and Analysis, "2015 Motor Vehicle Crashes: Overview," Traffic Safety Facts Research Note, National Highway Traffic Safety Administration, Report No. DOT HS 812 318, August 2016.
4. Fagnant, D. and Kockelman K., "Preparing a Nation for Autonomous Vehicles: Opportunities, Barriers and Policy Recommendations," *Transportation Research Part A: Policy and Practice* 77, no. 2015 (2015): 167-181, doi:10.1016/j.tra.2015.04.003.
5. Blincoe, L., Seay, A., Zaloshnja, E., Miller, T. et al., "The Economic Impact of Motor Vehicle Crashes, 2000," National Highway Traffic Safety Administration, Report No. DOT HS 809 446, May 2016.
6. Fertuck, D., Feldmaier, R., and Hurtado, M., "The Technology of Automated and Connected Vehicles," accessed August 2015, http://www.matecnetworks.org/webinars/pdf/CAAT%20Webinar%20August%202015%20with%20Valeo%20Final_8-26-15.pdf.
7. Litman, T., "Autonomous Vehicle Implementation Predictions: Implications for Transport Planning," presentation at *2015 Transportation Research Board Annual Meeting*, January 2015.
8. Casner, S., Hutchins, E., and Norman D., "The Challenges of Partially Automated Driving," *Communications of the ACM Magazine* 59, 5 (2016): 70-77, doi:10.1145/2830565.
9. Foundation for Traffic Safety, "Aggressive Driving: Research Update," accessed April 2009, https://www.aaafoundation.org/sites/default/files/AggressiveDriving ResearchUpdate2009.pdf.
10. Tasca, L., "A Review of the Literature on Aggressive Driving Research," *Aggressive Driving Issues Conference*, 2000.
11. Nillson, J., "Safe Transitions to Manual Driving from Faulty Automated Driving System," Ph.D. thesis, Department of Signals and Systems, Mechatronics, Chalmers University of Technology, Göteborg, 2014.
12. International Organization for Standardization, "Transport Information and Control Systems—Adaptive Cruise Control Systems—Performance Requirements and Test Procedures," ISO Standard 15622, April 2010.
13. Avizienis, A., Laprie, J., Randell, B., and Landwehr, C., "Basic Concepts and Taxonomy of Dependable and Secure Computing," *IEEE Transactions on Dependable and Secure Computing* 1, no. 1 (2004): 11-33, doi:10.1109/TDSC.2004.2.

14. Spooner, J. and Passino, K., "Fault-Tolerant Control for Automated Highway Systems," *IEEE Transactions on Vehicular Technology* 46, no. 3 (1997): 770-785, doi:10.1109/25.618202.

15. Kalra, N. and Paddock, S., "Driving to Safety: How Many Miles of Driving Would It Take to Demonstrate Autonomous Vehicle Reliability?" RAND Corporation, 2016, doi:10.7249/rr1478.

16. Waeltermann, P., "Hardware-in-the-Loop: The Technology for Testing Electronic Controls in Automotive Engineering," *6th Paderborn Workshop on Designing Mechatronic Systems*, 2009.

17. Joshi, A., "Real-Time Implementation and Validation for Automated Path Following Lateral Control Using Hardware- in-the-Loop (HIL) Simulation," SAE Technical Paper 2017-01-1683, 2017, doi:10.4271/2017-01-1683.

18. Khan, J., "A Standardized Process Flow for Creating and Maintaining Component Level Hardware in the Loop Simulation Test Bench," SAE Technical Paper 2016-01-0052, 2016, doi:10.4271/2016-01-0052.

19. Lamberg, K. and Waeltermann P., "Using HIL Simulation to Test Mechatronic Components in Automotive Engineering," *2nd Congress on Mechatronics in Automobiles*, November 2000.

20. Chan, B., "Framework for Developing Active Safety System for Commercial Vehicles," presentation at *2016 dSPACE Technology Conference*, September 2016.

21. Kohl, S., Lamberg, K., Otterbach, R., and Stroop, J., "Simulation, Implementation and Testing of Distributed, Time- Controlled Vehicle Systems," *IAV Symposium*, 2003.

22. Otterbach, R. and Kohl, S., "Efficient Function and Software Development," *2nd Paderborn Workshop on Intelligent Mechatronic Systems*, 2004.

23. Freund, E. and Mayr R., "Nonlinear Path Control in Automated Vehicle Guidance," *IEEE Transactions on Robotics and Automation* 13, no. 1 (1997): 49-60, doi:10.1109/70.554346.

24. Sha, L., "Using Simplicity to Control Complexity," *IEEE Software* 18, no. 4 (2001): 20-28, doi:10.1109/MS.2001.936213.

25. Boran, C., "Bringing Autonomous Vehicles into Production: An Automotive OEM Perspective," *3rd Automated Vehicles Symposium*, 2016.

26. Shen, S., "Quantifying Drivers' Responses to Failures of Semi- Autonomous Vehicle Systems," Ph.D. thesis, Department of Industrial Engineering, Clemson University, Clemson, 2016.

27. Clark, H. and Feng, J., "Semi-Autonomous Vehicles: A Closer Look at the Take-Over," *2015 International Annual Meeting of the Human Factors and Ergonomics Society*, 2015.

28. Trimble, T., Bishop, R., Morgan, J., and Blanco, M., "Human Factors Evaluation of Level 2 and Level 3 Automated Driving Concepts: Past Research, State of Automation Technology, and Emerging System Concepts," National Highway Traffic Safety Administration, Report No. DOT HS 812 043, July 2014.

29. National Highway Traffic Safety Administration, "Preliminary Statement of Policy Concerning Automated Vehicles," accessed May 2013, http://www.nhtsa.gov/staticfiles/rulemaking/pdf/Automated_Vehicles_Policy.pdf.

30. Norman, D., "The 'Problem' with Automation: Inappropriate Feedback and Interaction, not 'Over-Automation'," *Philosophical Transactions of the Royal Society of London, Series B, Biological Sciences* 327, no. 1241 (1990): 585-593.

31. Sarter, N. and Woods, D., "How in the World Did We Ever Get into That Mode? Mode Error and Awareness in Supervisory Control," *Human Factors: The Journal of the Human Factors and Ergonomics Society* 37, no. 1 (1995): 5-19.

32. Ziegler, W., Franke, U., Renner, G., and Kühnle, A., "Computer Vision on the Road: A Lane Departure and Drowsy Driver Warning System," SAE Technical Paper 952256, 1995, doi:10.4271/952256.

33. Motoyama, S., Ohta, T., Watanabe, T., and Ito, Y., "Development of Lane Departure Warning System," *7th ITS World Congress*, 2000.

34. Suzuki, K. and Jansson, H., "An Analysis of Driver's Steering Behavior during Auditory or Haptic Warnings for the Designing of Lane Departure Warning System," *JSAE Review* 24, no. 1 (2003): 65-70, doi:10.1016/S0389-4304(02)00247-3.

35. Information Resources Management, *Software Design and Development: Concepts, Methodologies, Tools, and Applications* (IGI Global, 2013), 310-334, ISBN:1466643013.

36. Mechanical Simulation Corporation, "CarSim/TruckSim/BikeSim Real-Time Hardware in the Loop," http://www.ishinho.com/product/data/CarSimRT_Presentation.pdf.

CHAPTER

9

Vehicle-GIS Assistant Driving System for Real-time Safety Speed Warning on Mountain Roads

Keyu Qian, Gangfeng Tan, Renjie Zhou, Binyu Mei, and Wanyang Xia
Wuhan University of Technology

Downhill mountain roads are the accident prone sections because of their complexity and variety. Drivers rely more on driving experience and it is very easy to cause traffic accidents due to the negligence or the judgment failure. Traditional active safety systems, such as ABS, having subjecting to the driver's visual feedback, can't fully guarantee the downhill driving safety in complex terrain environments.

To enhance the safety of vehicles in the downhill, this study combines the characteristics of vehicle dynamics and the geographic information. Thus, through which the drivers could obtain the safety speed specified for his/her vehicle in the given downhill terrains and operate in advance to reduce traffic accidents due to driver's judgment failure and avoid the brake overheating and enhance the safety of vehicles in the downhill.

Firstly, the characteristics of the driving conditions under the change of various slope lengths and gradients are studied based on models of both engine braking and wheel braking. The real-time information and characteristics of the vehicle on the mountainous area, such as the weight, the speed, the gear and the engine RPM, the vehicle braking torque characteristics on the mountainous area are analyzed. Meanwhile, the electronic map and GPS positioning are utilized to calculate the road

CITATION: Qian, K., Tan, G., Zhou, R., Mei, B. et al., "Vehicle-GIS Assistant Driving System for Real-time Safety Speed Warning on Mountain Roads," SAE Technical Paper 2017-01-1400, 2017, doi:10.4271/2017-01-1400.

gradient information. Finally, geographic information, road traffic speed limits and the brake efficiency will be combined to obtain a suggested safe speed at the top of the downhill roads and give as safety speed pre-warning to the driver.

Simulation results and case studies the pre-warning scheme works well for downhill roads, especially for the length from 1 km to 1.5 km and the slope gradient ranged between 3% and 6%. It suggests that the proposed scheme is more effective for improvement of the safety driving in the hilly slope area, where the slope is not too steep and the slope length is not too long and can be widely used in such areas to protect the safety of downhill driving.

1. Introduction

Statistics show that 43% of the land in China are mountainous and hills. Due to natural restrictions, mountain roads usually contain steep and long downhill sections [1]. Over speed is one of the critical causes in traffic accidents. Haworth's study [2] points out that over speed is lethal in car accidents. The death caused by break failure when driving in mountain areas ranks high among all traffic accidents. What's more, perceptions merely based on driving experiences and naked eyes usually lead to erroneous judgment, which would lead to accidents. In all, it is of great significance to warn the driver safety speed proactively.

At present, many researches have been carried out on the safety of downhill running at home and abroad, which can be classified in three categories: design of truck escape ramps, improvement of vehicle brakes, and design of vehicle retarders.

1. In terms of road traffic, the researches on the safety of the ramp driving in foreign countries are mainly focused on the studies and design of the truck escape ramps.

FIGURE 1-1 Long downhill accidents in mountainous areas.

As early as 1956, the United States had carried out the design of the truck escape ramps. *A Policy on Geometric Design of Highway and Street* [3] formulated by AASHTO has special provisions for the setting of hedge lanes, providing systematized instruction for its design.

The research on the truck escape ramps started late in China, so there is no normative explanation to the design and setting of the truck escape ramps. Besides, the attitudes of the norms in China toward setting up truck escape ramps are obscure, so there is no detailed statement on the specific settings of the truck escape ramps. As a result, there exists a high risk factor in truck escape ramps in China, especially in those early built ones [4].

As an important measure to prevent brake failure, most of the escape lanes are capable of deviating errant vehicle from the main track, thus ensuring the safety of other vehicles. But some fail to guarantee the safety of the driver on the errant vehicle-minor incidents like scraping and cargo scattering to serious accidents such as disability or death happen now and then [5].

2. To prevent the brake temperature rises too high, since 1937 Block [6] the first to study the friction heat problem, many workers have done a lot of researches. The analysis of the temperature of the brake under different conditions is very important for the safety performance of the vehicle.

    In China's mountain motor transportation, it's common to use wheel drum water to cool the brake down so as to ensure sufficient braking efficiency. To achieve that, a water tank must be installed at a higher level and the water pressure stems from its gravitational potential energy is used to transport the water to the brake hub. However, this approach presents a significant security risk. First of all, it may destroy the frame structure. Secondly, asymmetric temperature distribution due to the nonuniformity of the shower water, causes extrusion stress in the main parts of the brake drum, brake shoe and so on which lead to deformation warping, crack, roundness loss, and even cracking of the ring. Thirdly, when the shower works, most part of the water it sprays out turned into vapor or drip on the road after cooling process, with a small part enters into the brake through the aperture created by the backing plate and the brake drum. As the drum brake internal space is basically sealed, when braking at a high temperature, the water turns into vapor and fills the chamber; when the brake gets cool; water vapors condenses into droplets and adhere to the surface of various parts, causing corrosion, or even rust [7].

FIGURE 1-2 The truck escape ramp.

3. Hydrodynamic retarder and eddy current retarder are commonly used to slow down the commercial vehicles, which are heavy and clumsy. The main body of the hydraulic retarder is a fixed wheel system with a plurality of blades [8]. The heat generated by the brake is rapidly and effectively dissipated by the vehicle cooling system without adversely affecting the surrounding components. However, the hydraulic retarder has the following drawbacks [8] [9]. First, low-speed braking performance remains to be improved; second, under non-braking conditions, the idle loss is to be reduced; third, it's difficult to manipulate. In addition, the hydraulic retarder also has some other disadvantages, such as complex manufacturing process, high cost, excessive heavy, etc.

   Eddy current retarder is proposed based on the eddy current effect, excited by the rotating magnetic field interaction of eddy current and magnetic field to generate braking torque. It is also imperfect for its large size, heavy body, large electricity consumption and its effect on the temperature of the surrounding environment. Therefore, there exists potential security risk in this method.

   In this paper, we proposed a scheme of vehicle speed pre-warning system which is based on the vehicle dynamic characteristics and digital map geographic information. Combined with the characteristics of downhill speed and temperature rise of the brake, the system will inform the driver the safety speed before the vehicle goes down the slope to achieve autonomous control of the driver, thereby reducing traffic accidents caused by driver's misstep and brake failure and improving the vehicle safety during downhill.

2. System Description

This section describes the system from three aspects, the fundamentals of the system, the basic structure of the system and the basic process of the system.

2.1 System Fundamentals

Road information such as the angle of the inclined plane, the length and designed speed limit of the slope is obtained in advance through the digital map. In the downhill, the car uses its own gravity as the driving force. Meanwhile, the main brake and the engine braking are used as the braking mode. The paper establishes the vehicle's own dynamic model and use the downhill road speed limit and brake temperature rise as constraint conditions to get the top slope safety speed. Finally, the system will alarm the driver the safety speed in advance. Therefore the driver can control the speed ahead of time in the purpose of reducing the traffic accident of the downhill section. The schematic diagram of the system is shown in Figures 2-1, 2-2.

2.2 The Basic Structure of the System

The gradient and length of the slope are obtained by extracting the road information from the digital map. The total mass is calculated based on the change of the tire press obtained by the pressure sensor. The information processing unit is a safe speed calculation module. Remind module is utilized to remind the driver of slope safety speed before the downhill.

FIGURE 2-1 Cause of the accident on the downhill section.

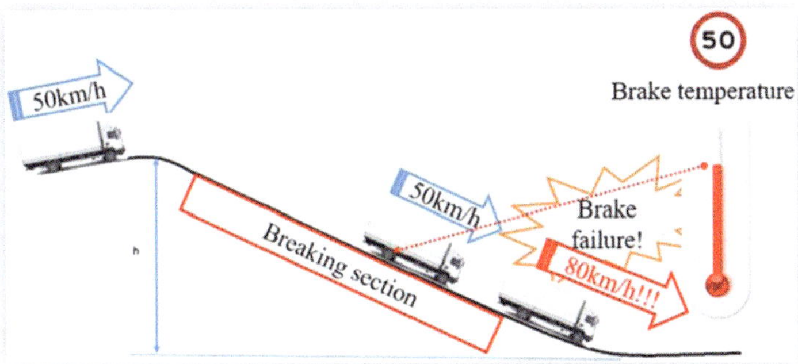

FIGURE 2-2 System working principle diagram.

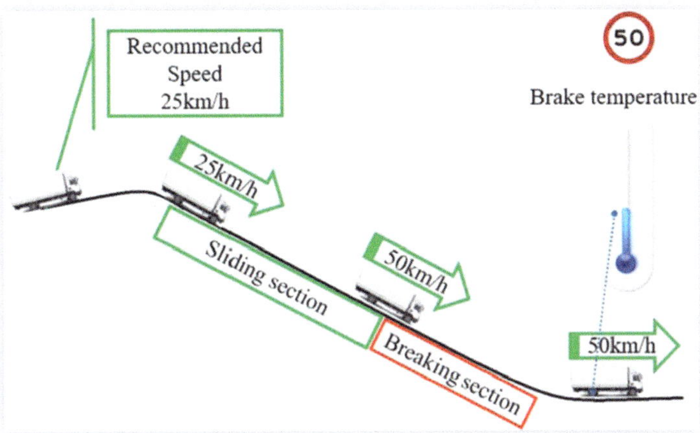

FIGURE 2-3 The basic structure of the system.

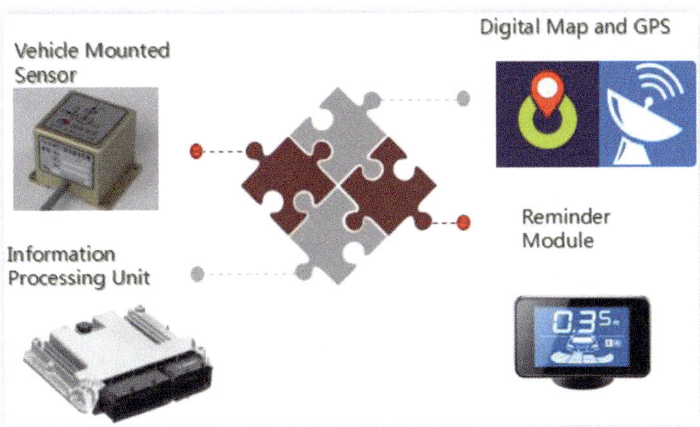

2.3 Basic Process of the System

Firstly, the scheme calls the digital map interface, and the road information is extracted to get the slope, slope length, traffic speed information of the front ramp. Secondly, the vehicle dynamics model and brake temperature rise model are established. Thirdly, this scheme uses the main brake temperature rise and traffic speed constraint to limit

FIGURE 2-4 Basic process of system.

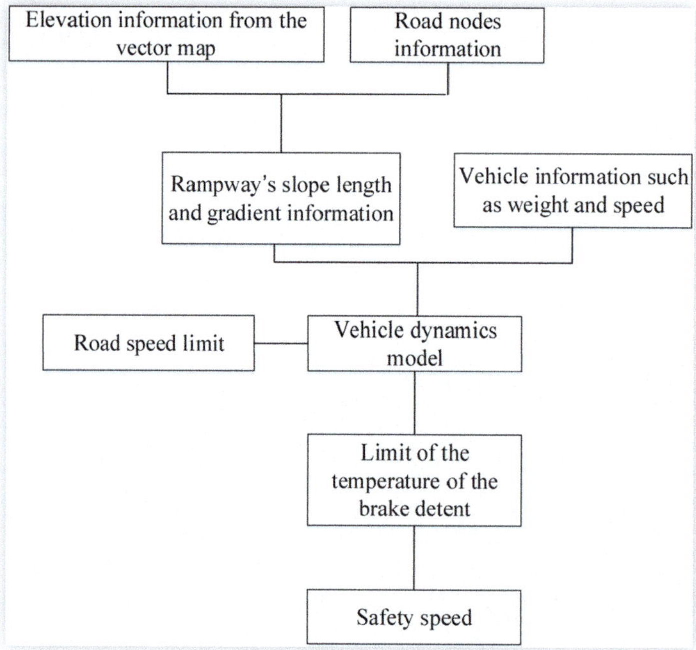

the top speed. Finally, these models are combined synthetically to obtain the safety speed of the top of the slope, and give the driving prompt to the driver ahead of the uphill slope.

3. Modeling and Constraints

The scheme proposed in this paper takes the temperature rise of the brake and the speed limit of the road as the constraint, and uses the vehicle dynamics to obtain the safety speed of the top of the slope. Therefore, this paper establishes the vehicle downhill braking model, vehicle weight model, brake temperature rise model. In the meantime, it takes the traffic speed limit in highway engineering technical standards into account.

3.1 Designed Highway Speed Limit in Mountainous Area

Design speed, also known as the calculation of traffic speed, refers to the maximum speed that could ensure the safety of medium skillful drivers(the driver who can follow the driver's operating procedures, can promptly discover and correctly judge the status of the vehicle being driven and use driving technology to ensure traffic safety)when the weather is good, the traffic isn't heavy and the condition of the road itself (like geometric elements, road, ancillary facilities, etc.) is the only thing to be considered during a drive. This design speed is often used as reference in the road design. The scheme takes road design speed as a constraint condition.

TABLE 3-1 Maximum slope length at different design speeds

slope length (m) slope(%)	Design speed (km/h)	120	100	80	60	40	30	20
3		900	1000	1100	1200			
4		700	800	900	1000	1100	1100	1200
5			600	700	800	900	900	1000
6				500	600	700	700	800
7						500	500	600
8						300	300	400
9							200	300
10								200

3.1.1 DESIGN SPEED OF DOWNHILL SECTION IN MOUNTAINOUS AREA

The research on the condition of road alignment under certain vehicle speed starts early in foreign countries and they presents limit value of longitudinal gradient incline angle and slope length under different designed speed which considered landform, climbing and brake performance of a certain vehicle model and driving behavior as well as psychological and physiological characteristic. The US AASHTO gives a sectionalized longitudinal gradient limit value based on typical truck speed. Japanese researches commutate the system by addition of gradient limit value in different climate region, such as snow area. Federal republic of Germany came up with the conclusion that slope sections are relatively dangerous after investigating domestic traffic accidents and that the number of accident grows significantly when the gradient reaches 6%. Longitudinal gradient in the country is designed to fit its landform.

In China's *Technical Standard of Highway Engineering* [10], the value of the slope length at different design speed is given, as shown in Table 3-1.

3.1.2 HIGHWAY SPEED LIMIT

According to China's *Technical Standard of Highway Engineering* [10], different levels of highway have different speed limits.

1. First grade highway as a national and provincial trunk road, through technical and economic feasibility studies, the design speed of which should adopt 100km / h or 80km / h. When it's taken as a large, medium-sized urban and rural areas with a large volume of traffic mixed traffic highway, the design speed can be 80km / h or 60km / h.
2. When the secondary highway is the trunk highway between the national and provincial trunk roads or the city, 80km / h can be used. The secondary highway is the distribution road with large mixed traffic volume in the urban-rural fringe. The design speed should be 60km / h.
3. When the third-class highway is taken as the feeder road, it should adopt 40km / h; in the sections with complex natural conditions such as terrain and geology, the design speed can be 30km / h.
4. Terrain, geology and other natural conditions of complex mountainous areas, or areas whose traffic is very small sections, can be used as forth-class highway whose design speed should be 20km /h.

3.2 Vehicle Dynamics Modeling of Commercial Vehicles on Downhill

In order to analyze the braking condition when the vehicle is running downhill, the model of the ramp braking is established. The real-time quality of the vehicle affects the safety speed of the top of the slope and can not be obtained directly by the sensor, so the vehicle weight model is established.

3.2.1 RAMP BRAKE MODEL

When the commercial vehicle is running downhill, it is mainly controlled by gravity, support force, braking force, rolling resistance and air resistance. The force is shown in the Figure 3-1.

1. The rolling resistance can be expressed by the following formula:

$$F_f = Gf\cos\alpha \tag{1}$$

2. Air resistance can be expressed as follows:

$$F_w = \frac{1}{2} C_D A_\rho V^2 \tag{2}$$

3. The ramp resistance can be expressed as follows:

$$F_i = G\sin\alpha \tag{3}$$

4. Ground braking [11] force can be expressed as follows:

$$F_x = \frac{T_\mu}{r} \tag{4}$$

$$T_\mu = k_b p_c + I_w w \tag{5}$$

5. Analysis of Engine Auxiliary Braking

The propulsion produced by the engine reaches the drive wheels through a series of power transmission devices. The power transmission mechanism between the engine and the driving wheel is called the automobile transmission system, which is mainly composed of the clutch, the transmission, the transmission shaft, the main reducer, the differential, and the half shaft. The engine output power would first go through the

FIGURE 3-1 Vehicle downhill force diagram.

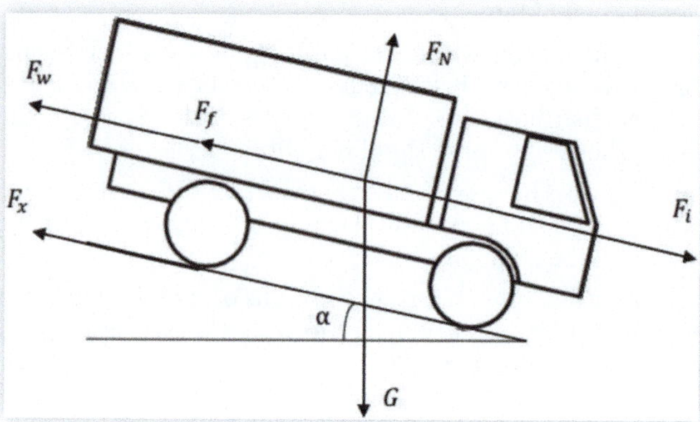

FIGURE 3-2 Engine power transmission line.

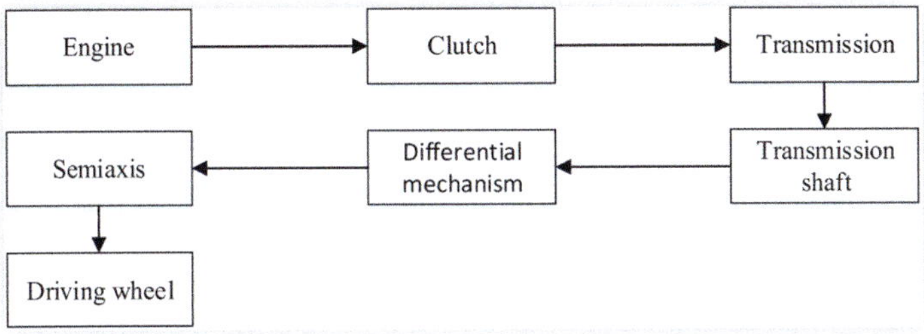

clutch, by the transmission torque and variable speed, the transmission shaft to the power transmission to the main reducer, and finally through the differential and the axle to drive the power to the drive wheel. The engine power transmission line is shown in Figure 3-2.

When the engine is braking, the braking power W_r is composed of 2 parts. One part is the heat loss in the engine gas service, that is, the work of the piston in the whole working cycle and the irreversible loss in the working process. Another part is the engine's mechanical losses which includes loss of friction between moving parts, engine driven accessories (such as gas distribution mechanism, generators, pumps etc.) power loss, which is denoted by W_f.

$$W_r = I\oint PdV + W_f \tag{6}$$

Here, the braking work caused by the mechanical loss can be expressed by the average friction pressure P_m, said that the above formula can be expressed as

$$W_r = I\oint PdV + \frac{\pi}{4} IP_m D^2 s \tag{7}$$

Due to the engine braking, the main source of brake power is the mechanical loss of the engine W_f [12].

So the simplified model is

$$W_f = \frac{\pi}{4} IP_m D^2 s \tag{8}$$

Because the average friction pressure is inevitably affected by the loss of the pump gas during the engine inlet and exhaust, it is difficult to obtain the exact value, which is usually calculated by the empirical formula. Usually it can be calculated by the following empirical formula of Ricardo in the United Kingdom:

$$P_m = 0.00697a + 0.00483\frac{n}{1000} + 0.0401\left(\frac{C_m}{10}\right)^2 MPa \tag{9}$$

a = ε (Swirl chamber diesel engine)
a = ε - 4 (Direct injection diesel engine)

3.2.2 WEIGHT MODEL

For a particular tire, relation between the internal air pressure changes and its radial load is approximately linear. Based on this relationship, use the tire pressure to measure vehicle load [13].

When a tire is pumped up, it shows the maximum volume under present condition as the compressed air always fills the space to the utmost. The shape of the tire changes

FIGURE 3-3 Model coordinate system establishment.

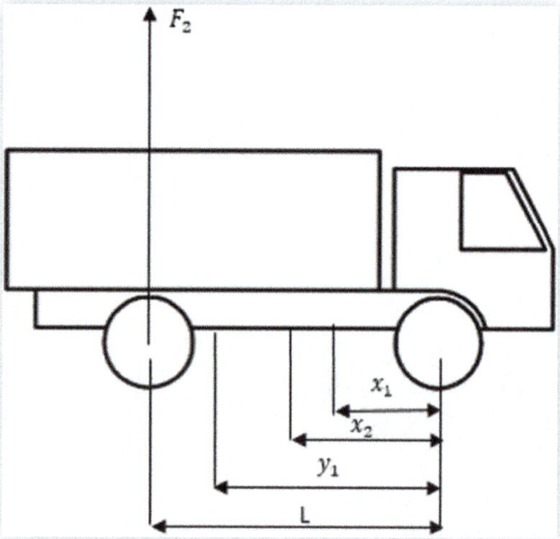

as the car is loaded and the inner pressure rises to bear the weight. Define volume reduction as ΔV. According to the gas isothermal compression theory:

$$P_1V_1 = (P_1 + \Delta P)(V_1 - \Delta V) \tag{10}$$

It can be known that the pressure of the compressed gas of the inner tube is increased by

$$\Delta \mathrm{P} = \frac{P_1 \Delta V}{(V_1 - \Delta V)} \tag{11}$$

Because ΔV is very small, the above equation can be approximated as

$$\Delta \mathrm{P} = \frac{P_1 \Delta V}{V_1} \tag{12}$$

ΔP is basically proportional to ΔV. Through the above-mentioned theory, the load-dependent ΔP is obtained to indirectly measure the load of the vehicle.

To make the car model as simple as possible, abstracted the 4 wheel truck into an ideal plane physical model, and make the following reasonable assumptions:

1. Truck body is symmetrical, the force of tire on both sides of the left and right are equal.
2. The center of gravity of the weight inside the compartment is the center of the compartment.

Take the front wheel center as the origin and establish the coordinate system as shown in Figure 3-3.

x_1--the position of the center of gravity of an automobile when it is empty, m
y_1--cargo center of gravity, m
x_2--position of center of gravity after load, m

According to the theorem of centroid motion:

$$x_2 = (x_1 m + y_2 M) \cdot (m + M)$$

Establish the torque balance formula:

$$x_2 \cdot (m + M) = F_2 L$$

So

$$F_2 = (m x_1 + M y_1) g / L$$

The pressure of a rear wheel:

$$F = F_2 / 2$$

M--quality of goods, kg
m--no - load car quality, kg

At the same time, assume that the tire inner tube is a regular ring with a circular cross section. The force surface is round after loading. The inner tyre stress surface is simplified into a cross section whose diameter is 1. The outer diameter of the tyre is R and the inner diameter is r.

The load volume changes of tyre before and after loading:

$$\Delta V = -\frac{2\pi}{3}\left(R - \sqrt{R^2 - l^2/4}\right)^3 \tag{13}$$

The L is used as the independent variable to start with the Taylor formula, and ignore the higher order of four times.

$$\Delta V = -2\pi l^6/3 \times 8^3 R^3 \tag{14}$$

Comprehensive above, can get the relationship between M and the pressure P can be obtained.

$$M = \frac{2(P_0 + \Delta P)RL^2\sqrt{\frac{12\pi^2 V_0 \Delta P}{(P + \Delta P)}/4 - x_{2m}}}{y_2 g} \tag{15}$$

As can be seen from the above equation, as long as the pressure sensor to measure the car when the tire pressure P_0 and tire pressure after the load P, we can get the car load. Other multi-wheeled commercial vehicles can be measured in the same way.

3.3 Brake Temperature Rise Model

Brake temperature rise is one of the constraints of the safety speed at the slope top. The temperature rise of the brake can be obtained by analyzing the energy conversion at the downhill of the vehicle.

3.3.1 ENERGY ANALYSIS OF DOWNHILL SECTION

Gravity potential

When the vehicle is running on the ramp, the elevation changes, thus the gravity potential energy is changed. The change of gravity potential energy is defined as E_P

$$E_P = G\Delta H \tag{16}$$

The change of kinetic energy

The kinetic energy change of the vehicle during descent is defined as E_K

$$E_k = \frac{1}{2}MV_2^2 - \frac{1}{2}MV_1^2 \tag{17}$$

The vehicle overcomes rolling resistance and wind resistance

When the vehicle is running on the ramp, it has to overcome the rolling resistance and air resistance. The work generated when overcoming the friction is:

$$W_{resistance} = \left[Gf\cos\alpha + \frac{1}{2}C_D A\rho\bar{V}^2\right] \cdot L \tag{18}$$

Work of auxiliary engine operation

$$W_{engine} = \frac{\pi}{4} IP_m D^2 S \tag{19}$$

CHAPTER 9

Brakes work

$$W_{brake} = F_x L \tag{20}$$

Energy analysis model

Based on the above formula, according to kinetic energy theorem, get:

$$\Delta E_K = \Delta E_P - W_{resistance} - W_{engine} - W_{brake} \tag{21}$$

3.3.2 ANALYSIS OF BRAKE TEMPERATURE RISE

When going downhill, the vehicle's gravitational potential energy will convert into kinetic energy. Accelerometer accrues to velocity. Constant braking force is required when going down a long slope so as to make sure that the vehicle still processes a braking property when it reaches the bottom. The heat generated in the brake during constant braking process decent the friction moment. The friction moment descents significantly when the temperature reaches 300°C, namely the "heat fading" phenomenon. When it reaches 600, the braking capacity may fall to 0[14]. The braking efficiency of the brake is high when the brake temperature is controlled at 200 ° C [15].

The total energy consumption of the main brake when the vehicle is going downhill can be obtained from the following formula:

$$W_{brake} = F_x L = G\sin\alpha L + \frac{1}{2}MV_1^2 - \frac{1}{2}MV_2^2 - [\mathrm{Gf}\cos\alpha + \frac{1}{2}C_D \mathrm{A}\rho\bar{V}^2]\cdot \mathrm{L} - \frac{\pi}{4}IP_m D^2 S \tag{22}$$

According to theoretical researches, about 95% of the heat generated during braking would be absolved by drum brake or brake disc, with rest of the heat be absolved by fraction plat or brake pad. Thus, it is supposed in this paper that the heat is all absolved by drum brake and we won't discuss about the fraction absolved by the brake shoe and brake system elements. When the drum brake works, heat will be generated between brake shoe linings and the internal surface of brake drum. A certain part of the heat is dispersed through transmitting, convection and radiation, while the rest accumulates within the brake drum and brake shoe which raise the temperature of the brake drum.

Without considering the heat conduction, the heat radiation and the heat convection, the W_{brake} is the total heat dissipation Q_t of the brake.

According to the thermodynamic formula

$$\Delta Q_t = mc\Delta T \tag{23}$$

We can get the brake temperature rise.

4. Case Studies and Simulations

Based on the above-mentioned model, the relationship between the safe speed of slope top and slope length, slope gradient and vehicle weight is studied. This paper simulates the safety speed at the slope top under the condition of different ramps and vehicle weight in MATLAB.

TABLE 4-1 Basic vehicle information

Project	Unit	Indications and Parameters
Wheelbase	mm	5600
Body length	m	9.995
Body width	m	2.5
Body height	m	3.96
Quality(full load)	t	14.49
Torque	N· m	800
Tire specifications	-	100R20
Number of tires	-	6
Specific heat capacity of brake	J/kg°C	460
Brake quality	kg	80
Bore × stroke	mm^2	124×107
Rolling resistance coefficient	-	0.010
Compression ratio	-	17.3
Air resistance coefficient × windward area	m^2	$C_DA = 3.271$

4.1 The Main Parameters of the Target Vehicle Model

In view of Chinese common truck model, performance and load, the paper selects the Dongfeng Tianjin Truck DFL5140XXYB4 as the test vehicle which is equipped by Dongfeng Cummins ISDe210 40 inter-cooling inline diesel engine (with 6 cylinders). The basic information is as followings.

This paper simulates downhill driving according to the driving habits of most drivers, supposing that the vehicle first keeps accelerating without brake when going down from the top and moves in constant speed when it reaches the highest speed limit. The safety speed obtained by this scheme can satisfies most drivers' driving habits and is more practical. It brakes only during the constant speed phase. The brake is most efficient at 200 °C, but when the temperature reaches 300 °C or even higher, its friction torque will decrease significantly which is called the "thermal recession" phenomenon brake of the brake. When the brake temperature exceeds 600 °C, the vehicle braking capacity may be lower to zero. Therefore, in the simulation, the brake temperature constrain is 200 °C. During the simulation, the formula (22) equals to the formula (23) and the temperature rise is under 200°C. The DFL5140XXYB4 truck is used as the test vehicle for simulation. MATLAB will serve as a simulation platform.

4.2 Safety Speed Simulation of Downhill Section

According to the *Technical Standard of Highway Engineering*[10] in China, the designed speed limit is 80km / h or 60km / h for first class highway, 60km / h or 40km / h for second class highway, 40km/h or 30km/h for third class highway and 20km/h for forth class highway. In order to obtain the corresponding safety speed at the top of the slope for a certain type of vehicle driving on highways of different class, do simulate tests of a full loaded vehicle going downhill from slopes of different length and gradient angle and compute the safety speed. In the simulation, gradient angle varies from 3% to 9% because they are common in Chinese mountainous area; the maximum slope length is 2500 meters

FIGURE 4-1 The safety speed at the slope top when the speed limit is 80km/h at the slope bottom.

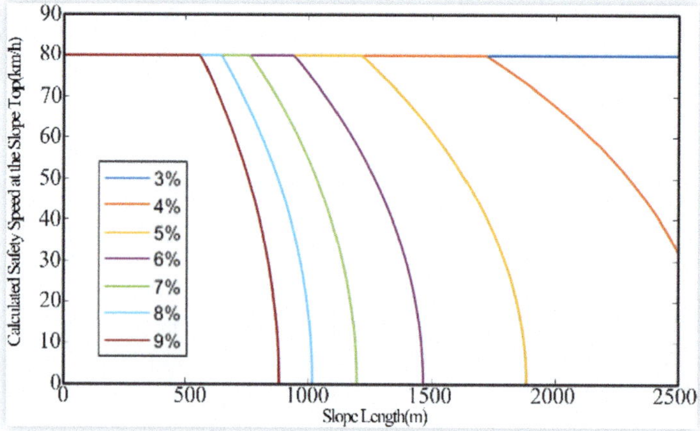

and brake temperature constrain is 200 °C because very long slope would lead to excessive brake load stems from the brake temperature rise and the auxiliary braking must be used.

Figure 4-1 shows the speed at the slope top under different gradient angle and the slope length when the slope speed limit is 80km / h. In the figure, the slope 3% line is a straight line parallel to the abscissa axis. The slope 4% line parallels to the abscissa axis and bends at point (1717, 80), then the speed gradually decreased. The slope 5% line bends at point (1214, 80) and the speed gradually decreases until the intersection with the abscissa, the intersection point is (1882, 0). The slope 6% line bends at point (929, 80), and the intersection of the axis of abscissa is (1432, 0). The slope 7% line bends at point (762, 80), and the intersection of the axis of abscissa is (1195, 0). The slope 8% line bends at point (646, 80), and the intersection of the axis of abscissa is (1015, 0). The slope 9% line bends at point (560, 80), and the intersection of the axis of abscissa is (881, 0). Table 4-2 shows the turning point and intersection point coordinates.

As can be seen from Figure 4-1, when the ramp limit speed is 80km/h, the brake temperature won't rise to over 200°C if vehicles drive under the speed limit on a slope which has 3% gradient angle and less than 2500 meters in length. The reason is that the slope gradient is gentle, gravity work is limited, meanwhile, the ramp resistance is small, so the braking force which is to keep the vehicle at a constant speed is small. Therefore, the safety speed can maintain the braking in the whole downhill not be disabled. The brake temperature would be under 200°C if the vehicle drives within the limit speed on a slope which has 4% gradient angle and less than 1717 meters long, but as the length exceeds 1717 meters, braking distance is longer, if the vehicle braking in the whole downhill, due

TABLE 4-2 Turning point and intersection point coordinates

Slope gradient	Turning point	Abscissa Intersection
3%	-	-
4%	(1717,80)	-
5%	(1214,80)	(1882,0)
6%	(939,80)	(1432,0)
7%	(762,80)	(1195,0)
8%	(646,80)	(1015,0)
9%	(560,80)	(881,0)

to the braking force to do a larger work, will cause the brake temperature rise higher than 200 °C. So the alarm safety speed at the slope top will decreases accordingly with the increase of slope length in order to ensure the efficiency of the brake. In this case, drivers who get top safety warning speed, will control the speed of the vehicle around it. In the downhill process, the vehicle gradually accelerates due to gravity, but the speed does not exceed the speed limit of the road. If the speed approaches the road speed limit, drivers can keep the speed under the limit through the brake control. The warning speed will effectively control the brake temperature rise, ensure the traffic safety. When the slope gradient is 5% and the slope length is greater than 1882 meters, because the slope gradient is bigger and the slope length is long, so the gravity work is bigger. Meanwhile, the braking force which is to keep the vehicle at a constant speed is bigger. Minimum braking force acting required to limit vehicle speed less than 80 km / h is greater than 200 ° C. The speed alarm is only helpful in slowing down the speed of temperature rise but not eliminating the danger. Therefore, other retarding device is in need to relieve the brake. When the slope is 6%, 7%, 8% and 9%, the analysis method is the same as 5% slope.

Figure 4-2 to Figure 4-5 is the slope safety speed of the slope with different gradient angle and length under speed limit of 60km / h, 40km / h, 30km / h and 20km / h.

FIGURE 4-2 The safety speed at the slope top when the speed limit is 60km/h at the slope bottom.

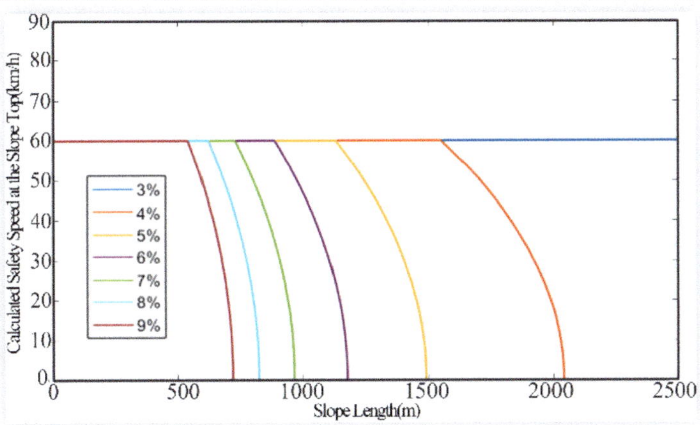

FIGURE 4-3 The safety speed at the slope top when the speed limit is 40km/h at the slope bottom.

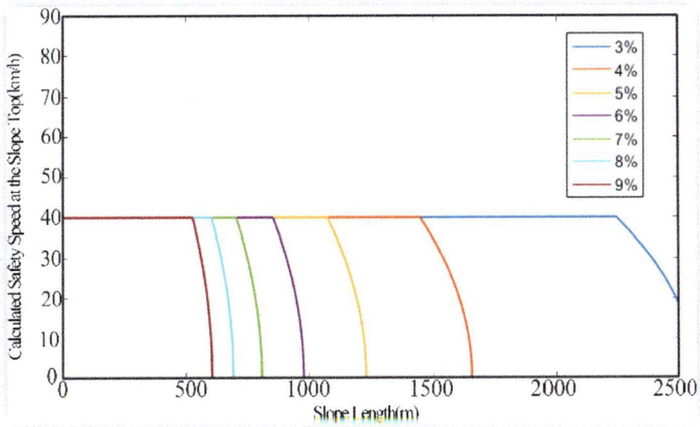

CHAPTER 9

FIGURE 4-4 The safety speed at the slope top when the speed limit is30km/h at the slope bottom.

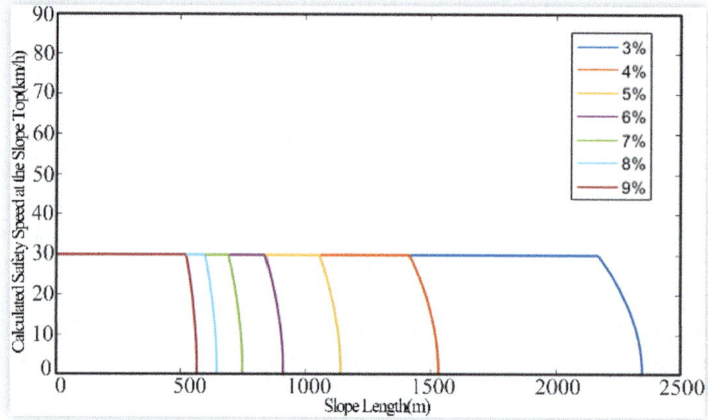

FIGURE 4-5 The safety speed at the slope top when the speed limit is 20km/h at the slope bottom.

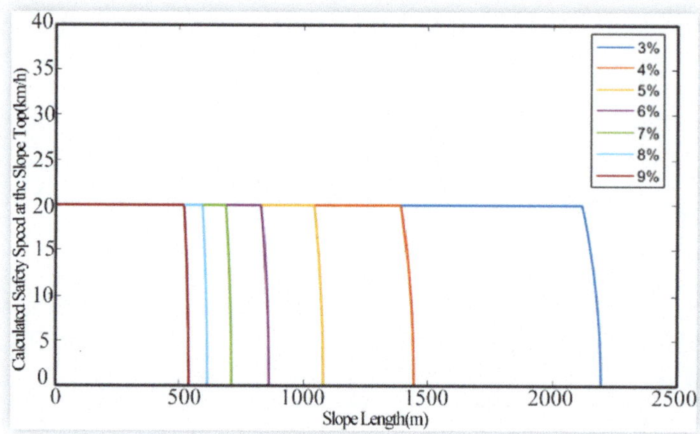

The analysis method is the same as Figure 4-1. From the figure, when the slope gradient angle is confirmed, there exists a critical slope length and a critical braking distance. When the length of the slope is less than the critical slope, the braking distance is less than the critical braking distance, the braking force which ensures the vehicle speed lower than the safety speed is small, so the brake temperature rise will not exceed the critical temperature of 200 °C, braking efficiency of the brake is high. When the slope length is longer than the critical slope length, the safety speed indicated at the top of the slope will be correspondingly reduced to reduce the braking force work. When the slope length is too long, to ensure the vehicle speed is lower than the safety speed must ensure that a longer braking distance and the braking force acting more. The vehicle did not reach the bottom of the slope when the brake temperature rises up to 200°C. So other auxiliary braking devices are required. The coordinates of the turning point and the coordinate of the abscissa intersection when the road speed is 60 km / h to 20 km / h are shown in Table 4-3.

And Figures 4-1 to 4-5 shows that when the road limit speed is reduced, because the average speed decreases, the wind resistance reduces, the required braking force

TABLE 4-3 Turning point and intersection point coordinates

Road speed limit(km/h)	The gradient of the slope (%)	Turning point	Abscissa Intersection
60	3	(2498,60)	-
	4	(1548,60)	(2038,0)
	5	(1127,60)	(1489,0)
	6	(886,60)	(1173,0)
	7	(727,60)	(964,0)
	8	(620,60)	(823,0)
	9	(540,60)	(657,0)
40	3	(2244,40)	-
	4	(1446,40)	(1656,0)
	5	(1072,40)	(1229,0)
	6	(851,40)	(976,0)
	7	(704,40)	(807,0)
	8	(603,40)	(691,0)
	9	(527,40)	(604,0)
30	3	(2166,30)	(2344,0)
	4	(1414,30)	(1531,0)
	5	(1054.30)	(1141,0)
	6	(840,30)	(909,0)
	7	(696,30)	(753,0)
	8	(598,30)	(647,0)
	9	(522,30)	(565,0)
20	3	(2114,20)	(2192,0)
	4	(1391,20)	(1442,0)
	5	(1041,20)	(1079,0)
	6	(832,20)	(862,0)
	7	(691,20)	(716,0)
	8	(593,20)	(614,0)
	9	(519,20)	(538,0)

increases, the critical braking distance decreases, critical slope which can make the vehicle follow the road speed limit of the whole length becomes smaller, the brake burden increases, the maximum slope length when the brake temperature reaches 200°C is smaller, too. It can be seen from the figure that at the middle slope length, the safety speed obtained by the velocity warning method can effectively reduce the excessive temperature rise of the brake. However, when the slope length is too long or the slope is too steep, only the system will not be able to fully protect traffic safety, vehicles need to use other auxiliary braking system.

4.3 Influence of Vehicle Weight on Slope Top Safety Speed

Slope gradient, slope length and vehicle real-time gravity are the factors influencing the safety speed of slope top. In order to determine the effect of vehicle weight on

FIGURE 4-6 Influence of vehicle weight on safety speed.

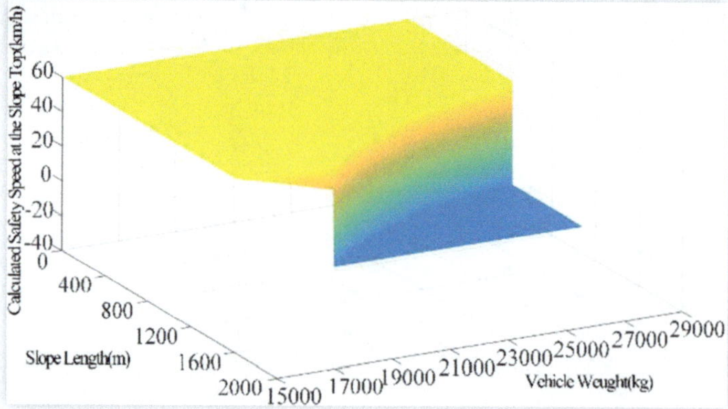

the safety speed of the top of the slope, the change of vehicle weight is introduced. The total mileage of China's highway is 4.463 million kilometers, which the highway accounts for 2.5%, the first-class highway accounts for 1.9%, the second-class highway accounts for 7.8% and the third-class highway accounts for 6.3%. Therefore take the secondary road as a simulation road, that is, the speed limit is 60km/h. According to *Study on Distribution Law of Slope Land in Mountainous Area of Northern China*[16], the slopes of 2 ° -6 ° accounts for 38.23% of the total mountain area, and 6 ° -15 ° accounts for 28.69% of the total area. So take the gradient of 5%, namely 2.86°, slope as simulation of road conditions. When the slope length is too long, the safety speed warning only plays a role in slowing down the rising speed of the brake temperature and cannot completely eliminate the danger. Combining with the previous analysis, take the middle slope length 2000 m as the maximum simulation slope length. Get Figure 4-6.

In Figure 4-6, when the slope length is short, the weight is less, the graph is a plane parallel to the abscissa axis, the greater the vehicle weight, the shorter the corresponding length of the slope. With the increase of speed and weight of the vehicle, the calculated safety speed at the top of the slope is reduced gradually. The greater the vehicle weight is, the more obvious the decreasing trend of the safety speed at the top of the slope. When the slope length and vehicle weight reaches a certain value, the speed becomes 0.The reason is that when the slope length and weight are small, gravity work is small, the temperature of the brake into the energy is also small, the temperature rise is less than 200 °C. And when the slope lengths are equal, the larger the vehicle weight, the greater the required braking force to maintain uniform, so the maximum braking distance when the brake reaches 200 °C is shorter. With the increase of vehicle weight, when the same braking method is adopted under the same gradient and slope length, the thermal energy converted from gravitational potential energy to brake can be reduced only by reducing the brake strength. Therefore, the speed is reduced accordingly. When the slope length and weight is too large, because of gravity into the brake temperature rise of the energy is too large, the system can only play the role of reducing the temperature rise, auxiliary braking devices are needed to ensure traffic safety. For example, when the vehicle weight is greater than 800kg, slope length is greater than 2100 meters, the system prompts the vehicle speed only plays a role in reducing the temperature rise of the brake, and can not fully guarantee traffic safety. At this point the auxiliary driving system is need to share the brake load, reduce the brake temperature rise.

4.4 Brake Temperature Rise Comparison

Brake temperature rise before and after using this speed warning scheme is compared in order to verify the effectiveness of the system. It is assumed that the driver will travel at the road speed limit before using the warning speed, and obtain the comparison chart of brake temperature rise under different road speed limits.

Figure 4-7 is the brake temperature rise contrast diagram when the ramp limit speed is 80km/h. The figure shows that the road speed limit of 80km / h, slope length of 0-2500 meters, when the slope of 3%, the brake temperature will not rise to exceed 200 °C. When the gradient is 4%, the intersection point is (1717, 200), indicating that if the slope length is more than 1717 meters, if the vehicle is still moving at the speed limit, the brake temperature will be too high and lead to brake efficiency or brake failure. And the safety speed given by the system can keep brake temperature maintaining at about 200 °C, so as to ensure traffic safety. Similarly, the intersection coordinates are (1214,200) (939,200) (762,200) (646,200) (560,200)

when the gradient is 5%, 6%, 7%, 8% and 9%,indicating that when the slope length is longer than the critical slope length of the corresponding slope, the brake temperature rise will be high, the slope length after accumulation is likely to cause the brake temperature rise too high and dangerous, this time using the safety slope speed can effectively reduce the brake failure risks.

For the study on the driving distance change before and after using the system when the brake temperature rise limit of 200°C, select the slope bottom limit speed of 80km/h, the slope gradient from 3% to 6% for analysis. From Figure 4-8 we can know, when the gradient is 3% to 9%, before using the scheme, the vehicle can only drive to 2615, 1717, 1214, 939, 762, 646, 560 meters after the brake temperature reaches 200°C, and after using this scheme, the vehicle can travel to 2966, 2499, 1882, 1453, 1195, 1015, 881 meters after the brake temperature reaches 200°C. It can be known that the system can effectively improve the safety of the vehicle in the downhill.

FIGURE 4-7 Comparison of brake temperature rise when the speed limit is 80km/h at the slope bottom.

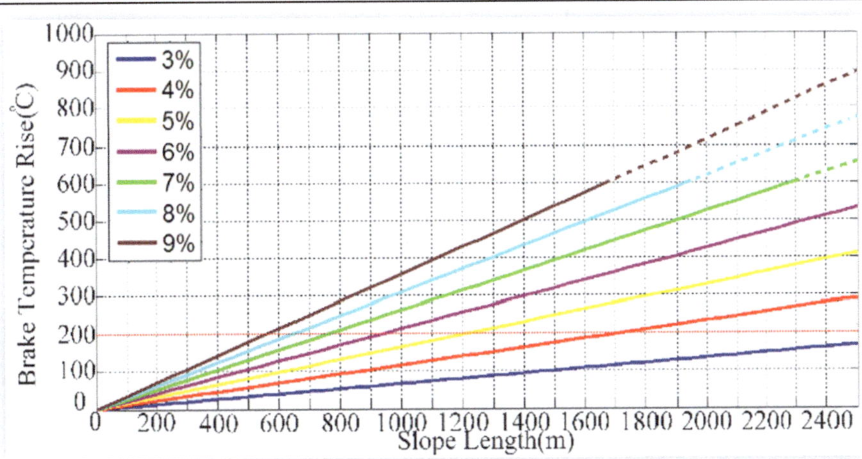

FIGURE 4-8 Travel distance comparison.

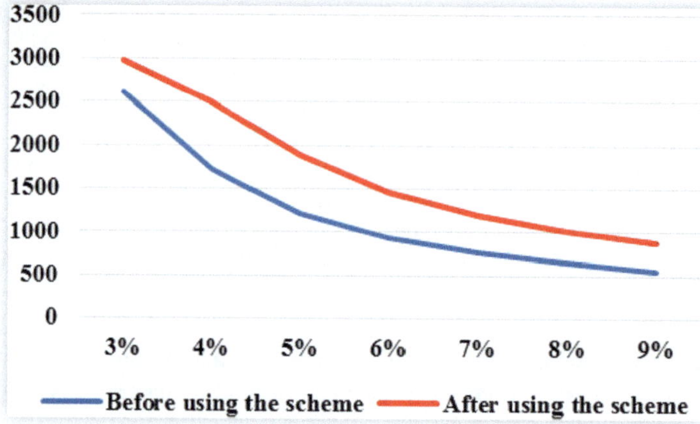

5. Conclusion

Based on the basic theory of vehicle driving dynamics, this paper establishes a propelling-braking model, a brake temperature rise model and a vehicle weight model on the mountain roads, and puts forward a kind of pre-warning scheme for different commercial vehicles which will inform the driver in advance the safety speed to drive down the slope. The characteristics of the slope safety speed values under various slope gradient, slope length and speed limits have been studied by using MATLAB simulation. The conclusions are as drawn.

1. When the gradient angel ranges from 3% to 6% and the slope ranges from 500 to 1000 meters long, the scheme is of little practical significance. In the slope of 6% -9%, the scheme can only play a role in slowing down the temperature rise of the brake. In the slope of 3% -6%, slope length of 1000-1500 meters, when applied in such conditions, the scheme will reduce brake temperature by about 50% and effectively reduce the accidents cause by the brake failure. Besides, this kind of slope is common in China's mountainous areas so the scheme can be widely applied.
2. The computation of the safety speed in this paper takes vehicle weight, specific road conditions into consideration, making the calculated speed more normative and rational. This scheme will provide a more exclusive safety speed warning, which is of great significance in reducing the traffic accidents in the downhill sections.
3. In terms of speed pre-warning, the system combines the digital map which provides gradient and slope length to be invoked ahead of the downhill section. The safety speed of the top will be computed and be announced to the driver, so the driver can actively control the speed. The system thus contributes a lot to improve driving safety in downhill sections.

References

1. Liying, S., *The Driver Visual speed controlling Design Method of the Long Steep Downgrade Considering the Temperature Rise of Drum Break* (Wuhan University of Technology, 2010).
2. Haworth, N. and Rechnitzer, G., "Description of fatal crashes involving various casual variables (Report No CR 119)," Federal Office of Road Safety, Canberra,1993.

3. AASHTO Green Book: A Policy on Geometric Design of Highway and Street.
4. Li, S., *Design of Mountain Highway Emergency Lane* (Chang'an University, 2015).
5. Zhang, X., "Analysis on the Causes of Common Accidents and Design Defects in Safe Haven," *Gansu Science and Technology* 02 (2013): 74-76.
6. Block, H., "Theoretical Studies of Temperature Rise at Surfaces of Actual Contact Underoilness Lubricating Conditions," *Proceeding of General Discussion of Lubricants*, The Institute of Mechanical Engineer, 1937, 225-235.
7. Sheng, Y., "Safety Hidden Trouble of Air Pressure Water Spraying Device and its Improvement Measures," *Highways Automotive Applications* 02 (2007): 5-7.
8. Yan, J., *Research on High Power Hydraulic Retarder Design and Brake Torque Control Algorithm* (Wuhan University of Technology, 2010).
9. Guo, X. and Zheng, W., "Modern Vehicle Retarder, *Automobile Research & Development* 05 (2005): 39-42.
10. Construction of the highway Committee, *Technical Standard of Highway Engineering* (China Communications Press, 2014).
11. Pei, X., Liu, X., and Qi, Z., "Road Surface Identification Method for Vehicle Longitudinal Integrated Control System," *China Journal of Highway and Transport* 05 (2014): 177-182.
12. Yan, S., *Performance Simulation Research of Engine Supplementary Brake* (Hunan University, 2013).
13. Luo, G., Huang, L., and Shi, W., "Research on Measurement Technology of Static Load Based on Tire Pressure," *Acta Metrologica Sinica*, 32, no. 4 (2011): 333-337.
14. Gu, Y., *The Experimental Research on the Temperature Rise of Drum Brake* (Chang'an University, 2008)
15. Bowman Brain, L., "Grade Severity Rating System (GSRS)-Users Manual," Federal Highway Administration Office of Safety & Traffic Operation Research and Development, 1989.
16. Bai, X., Zhang, R., and Zhou, D., "Study on Distribution Law of Slope Land in Mountainous Area of Northern China," *Soil and Water Conservation in China* 05 (2010): 39-40.

Definitions/Abbreviations

F_x - Ground braking force.
F_f - Rolling resistance.
F_w - Air resistance
F_i - Ramp resistance.
F_N - Ground support for vehicles.
G - Commercial vehicle gravity.
α - Angle of slope.
f - Rolling resistance coefficient
C_D - Aerodynamic drag coefficient
ρ - Air density.
A - Windward area.
V - The speed of the vehicle.

T_μ - Braking torque.
r - Wheel radius.
k_b - Gain coefficient of the brake.
p_c - Wheel cylinder pressure.
I_w - Wheel inertia.
w - The angular acceleration of the wheel.
I - The number of cylinders.
P - Cylinder pressure.
V - Cylinder volume.
P_m - Average friction pressure.
D - Cylinder diameter.
S - Piston working stroke.
ε - Engine compression ratio
C_m - Piston average speed.
n - Engine speed.
L - Slope length.
V_1 - Slope top speed.
V_2 - Slope bottom speed.
$\bar{V}$ - Average speed during travel.
M - The total mass of vehicle.
m - Brake quality.
c - Specific heat capacity of brake.
ΔT - Brake temperature rise.

CHAPTER 10

Overtaking or Merging? Eco-Routing Decision and Speed Trajectory with Full Terrain Information

Ting Jun Li, Minghao Shen, and Hongyu Zheng
Jilin University

With vehicle platooning becoming an important research field in recent years, it is now imperative to introduce platoons as part of the dynamic environment, considering overtaking and merging possibilities. This article studies optimal speed trajectories and longitudinal control with optimized energy efficiency for an autonomous vehicle with several preceding platoons and full terrain information. It aims at improving the energy efficiency of vehicles with Advanced Driver Assistance Systems (ADAS). A forward discrete dynamic programming (DDP) algorithm with distance as the discretization basis is used to derive speed trajectories in the trade-off between air drag reduction and energy saved by utilizing the road slope information. The problem is decomposed into decisions whether to overtake or to merge into the nearest platoon with the assumption of sufficient distance among platoons. During the process, speed choices and cost function reflect interactions among the controlled vehicle, the platoons, and the road. Simulations confirm that the energy consumed can be reduced significantly when tracking optimal trajectories compared to driving by the existing model predictive control (MPC) tracking strategy and linear quadratic regulator (LQR). For a trip of 500 m, if the primary velocity of the ego vehicle is 7.5 m/s and the initial state of a preceding platoon is 10 m from the ego car, 8 m/s, the energy cost saving of the proposed solution can be up to 3.45%

CITATION: Li, T., Shen, M., and Zheng, H., "Overtaking or Merging? Eco-Routing Decision and Speed Trajectory with Full Terrain Information," SAE Technical Paper 2018-01-0038, 2018, doi:10.4271/2018-01-0038.

compared by LQR. When there are two platoons whose original states are 10 m, 7 m/s, and 35 m, 7.5 m/s respectively, the energy saving of the ego car with initial velocity of 8 m/s is 3.07% compared by LQR and 1.99% by MPC. Simulations indicate the potential of energy efficiency of the proposed method for further studies with more sophisticated conditions. Sensitivity analysis has shown that the energy saving is not sensitive to the initial speed of the ego car when it is relatively low.

Introduction

The areas of enhancing energy efficiency are of substantial interest to researchers. This is demonstrated by the number of research projects supported from different sources with a range of results (decision support systems/ surveillance systems, evaluation frameworks and guidelines, scientific reports etc.), as is said in [1]. Among all the approaches to address the problem, eco-driving is of great importance. There are two branches of related research projects, control and optimization, both leading to systematic solutions.

On one hand, the branch of control aims at deriving the optimal control allocation in relatively fixed scenarios. They range from torque distribution to adaptive cruise control (ACC) system and platoons. In [2, 3, 4, 5], Chen and Wang focus on acquiring the control allocation of electric ground vehicles(EGV) with DP method, adaptive energy-efficient control allocation (A-EECA), and a Karush-Kuhn-Tucker(KKT)-based algorithm. In [6, 7, 8, 9], S. Li, K. Li, et al. adopted MPC to address tracking issues, studied partial and full PnG driving modes in car-following scenarios, and considered the influence of road elevation information. In [10], S. Li presented an outline of platoon control techniques from the networked control angle, which breaks a platoon down into four correlative components. Rajamani, Tan [12] introduced an incorporated longitudinal and lateral control system for the operation of vehicles in platoons as part of the California PATH Program. On the other hand, the branch of optimization mostly weighs decisions, such as merging and overtaking, to obtain optimal scenarios. Liang discussed the break-even ratio for heavy duty trucks to decide whether to catch up and merge into a preceding platoon or not in [10]. Ajanović integrated the dynamic environment as constraints into a dynamic programming (DP) algorithm to work out the optimal trajectory tree in order to find out whether and where to overtake a preceding vehicle is most economical in [11]. Moreover, optimal trajectory tree and MPC-like replanning scheme are combined in the globally optimal MPC-like motion planning framework mentioned in [13] so that it can deal well with dynamic constraints.

Although optimal control in fixed scenarios has been studied in a lot of pieces of literature, decision optimization based on the dynamic environment is attracting more and more attention with the undergoing evaluation of electrification and informatization. Moreover, with platooning becoming so important an area, with projects in different countries on it [14], it is now imperative to take it into the consideration of the dynamic environment. This article aims at generating optimal speed trajectories and longitudinal control taking preceding platoons into account. It proposes an efficient way to handle the trade-off between the air reduction if merging and the energy saving by utilizing the terrain information if overtaking. Discrete dynamic programming (DDP) is used here, which has an inherent advantage of integrating the optimization and all kinds of constraints with discrete controls. Such a study helps to develop more thorough research projects into the dynamic environment and optimal trajectories in further studies. It may also contribute to enhancing the energy efficiency of the intelligent cities and highways.

The remainder of this paper is organized as follows. In section II, the vehicle dynamic model is represented. Optimization constraints are described in section III. In section IV, the DDP is illustrated as the motion planner. In section V, several comparisons of trajectories generated in MATLAB are presented to show the efficacy of the introduced energy-efficient algorithm. Conclusion are presented in section VI.

Vehicle Dynamic Model

The vehicle dynamic model can be defined with the following equations:

$$F_m = \frac{kT_m\eta}{r_w} \tag{1}$$

where F_m denotes the force transmitted to tires, k the transmission ratio of the car. T_m is the torque generated by the motor. η refers to the efficiency of the automotive driveline. r_w the radius of the wheel.

$$F_r = \frac{1}{2}\rho_d c_d A_f v^2 + c_r mg\cos\alpha + mg\sin\alpha \tag{2}$$

where F_r means the total resistance force. ρ_d refers to the density of the air. A_f means the frontal area of the autonomous car. The aerodynamic drag coefficient is c_d, while the rolling resistance coefficient c_r. Especially, m means the mass of the ego vehicle. The gravity acceleration is g and the road slope angle at present is α.

According to Newton's second law, the following criterion must hold:

$$ma = F_m - F_r \tag{3}$$

In this equation, a means its acceleration.

Besides, the angular velocity of tires ω can be calculated as:

$$\omega = \frac{iv}{r_w} \tag{4}$$

where i stands for the transmission ratio.

Optimization Constraints

The constraints are classified into two kinds, internal constraints and external ones.

Internal Constraints

For the internal constraints, ego vehicles themselves determine them, including ranges of velocity and torque. They are:

$$v_{\min} < v < v_{\max} \tag{5}$$

$$T_{\min} < T < T_{\max} \tag{6}$$

The $v_{\min}$ and $v_{\max}$ are the velocity minimum and the maximum. From $T_{\min}$ to $T_{\max}$ is the achievable range of torque that the actuator can provide.

For initial state:

$$v = v_0 \tag{7}$$

$$x = 0 \tag{8}$$

where x denotes the position where the ego car is at the present stage. The ego car refers to the vehicle that we intend to work out optimal speed trajectories for.

External Constraints

Considering the safety when overtaking and the comfort, some external constraints have to be set.

Platoons Initial State

We assume that the preceding platoons are moving in constant speed and scattered in abundant distances. Then we have:

$$\mathbf{X}_{\text{pl}} = \begin{bmatrix} x_{pl\_1} \\ x_{pl\_2} \\ \vdots \\ x_{pl\_n} \end{bmatrix} \tag{9}$$

and

$$\mathbf{V}_{\text{pl}} = \begin{bmatrix} v_{pl\_1} \\ v_{pl\_2} \\ \vdots \\ v_{pl\_n} \end{bmatrix} \tag{10}$$

In this equation, $\mathbf{X}_{\mathbf{pl}}$ and $\mathbf{V}_{\mathbf{pl}}$ are adopted to describe the initial state of n preceding platoons, where $x_{pl\_1}, x_{pl\_2}, \cdots, x_{pl\_n}$ refer to the initial distance between them and the ego car, and the $v_{pl\_1}, v_{pl\_2}, \cdots, v_{pl\_n}$ to the constant speed of platoons.

OVERTAKING SAFETY

The distance between the ego vehicle and preceding platoon i is defined as below:

$$\Delta x_{pl\_i} = x_{pl\_i} - x \tag{11}$$

where x denotes the position of the subject car.

Meanwhile, the overall state space of the distance $\mathbf{D}$ is defined as below,

$$\mathbf{D} = \begin{bmatrix} \Delta x_{pl\_1} \\ \Delta x_{pl\_2} \\ \vdots \\ \Delta x_{pl\_n} \end{bmatrix} \tag{12}$$

Obviously, $\mathbf{D}$ satisfy the equation provided below:

$$\mathbf{D} = \mathrm{X}_{\text{pl}} + \mathrm{V}_{\text{pl}} t - x \tag{13}$$

where t means the time consumed at the moment, and the s refers to the distance that the ego car has covered. Since we assume the distance between platoons are sufficient, we are able to figure out which platoon is nearest and how long the distance between them is simply from analyzing **D**.

In this way, we decompose the problem into several decisions on overtaking or merging into a single platoon. With the item in **D** with the smallest absolute value, named by Δx, we propose the overtaking process as below.

1. After the ego car drive into the distance of d_{pl} behind platoons, it begins to enjoy the air reduction below and is considered platooning. In this situation

$$c_d = c_d' \tag{14}$$

2. When the ego car drives into the distance of reparing,

$$d_{pp} < \Delta x < d_{pl} \tag{15}$$

If the biggest item of the predicted velocity state space v_{pr} in the next few steps can not satisfy what a safe overtaking maneuver need, the control system will set the torque so that the vehicle will maintain at the speed of the platoon. Namely, if

$$\max(v_{pr}) < v_{ot} \tag{16}$$

(where v_{ot} means safe overtaking speed and is defined in the next situation), then set

$$\max(v_{pr}) \leftarrow v_{\text{pl}} \tag{17}$$

In this case, the car will continue to platoon until there are some external factors interfering, for example, a downhill to help it accelerate.

3. In the overtaking rocess:

$$d_s < \Delta x < d_{pp} \tag{18}$$

To ensure the security when overtaking, the following criterion must hold

$$v_{ot} - v_{pl} \geq v_{safe} \tag{19}$$

where v_{safe} is defined manually and depends on l_0, the present traffic situation and other factors. Then it is demanded that the smallest item of the v_{pr} be maintained to be bigger than v_{ot} as below:

$$\min(v_{pr}) \geq v_{ot}$$

P.S. The maximum in the v_{pr} is bigger than the v_{ot} because of the preparing process.

Motion Planning Algorithm

Discrete Dynamic Programming (DP)

INTRODUCTION TO DP

Dynamic Programming (DP) was developed by Richard Bellman in 1963. It is a very strong method to solve optimal control problems. DP can handle all kinds of constraints, discrete controls, and none-differentiable dynamics [16]. It solves problems by solving subproblems first

FIGURE 1 The subscript pl, pp, and s stands for *platooning*, *preparing*, and driving aside for *security*, respectively. The platoon is positioned from D to E, and its length is l_0.

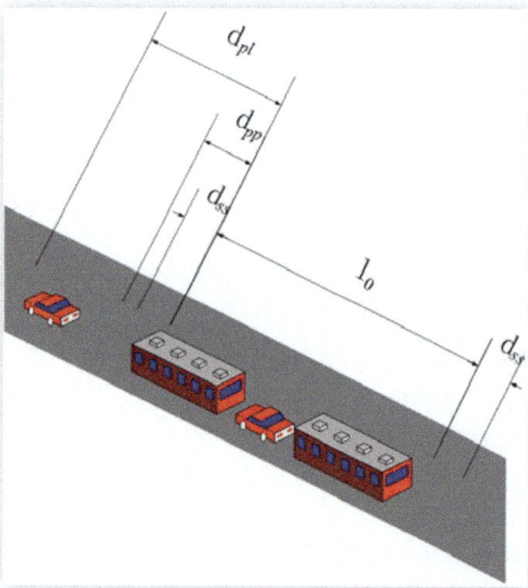

DP problem is generally divided into two parts, the iteration part, and the recursion part. In the first parts, we break up a problem into a series of overlapping subproblems, and then we solve each subproblem. For a typical DP problem, which is also called backward DP, the subproblem is if we choose control input u_i in stage k, which the optimal input u_j we should use in the next stage is among the attainable speed choices. For example, first we solve the tiniest subproblems, starting with the last transitions. In this situation, the current states are the final states, and the cost to the end is zero. We can calculate the cost of transitions from previous states to present states to learn the minimum cost to the end of states of them. Then we set the previous states to be the current ones and repeat the process. By this way, we obtain the optimal initial state and the knowledge of the minimum cost to the end of every state, which is called cost-to-go map.

The second part of the problems is a recursion part. Its idea is based on the optimal principal in [17]:

> *"An optimal policy has the property that whatever the initial state and initial decisions are, the remaining decisions must constitute an optimal policy with regard to the state resulting from the first decisions."*

Therefore, the DP problem, with the control inputs u_1, u_2, $\cdots$ u_n as solution,

$$J = \min_{u_1, u_2, \cdots u_n} f\left(u_1, u_2, \cdots u_n\right) \tag{21}$$

satisfy the equation:

$$J = \min_{u_1}\left(\min_{u_2, u_3, \cdots, u_n} f\left(u_2, u_3, \cdots, u_n\right)\right) \tag{22}$$

In this case, J denotes the cost of applying control inputs and being in the final state. Therefore it can generate optimal speed paths from the appropriate initial state by utilizing the cost-to-go map and go forward.

FORWARD DISCRETE DP

In this case, we need to calculate **D** in every stage. Therefore, the exact time that has passed since the initial stage is essential to the iteration part. Therefore a forward DP is used here. In the forward DP, a backward recursion follows a forward iteration. We access the optimal final state and the knowledge of the minimum cost from the start of every state, which is called cost-to-achieve map. Figure 3 is an example of one. It is derived from a forward DP with distance as stages and velocity as states. In this case, the initial speed of the ego car is 10 m/s, without platoons ahead.

As is said in [11], the practice of dynamic programming can use time or distance as the basis for discretization. Using distance is handful in looking for the energy optimal trajectory since the final time is not fixed. Therefore the x in the stage number $i + 1$ is

$$x_{i+1} = x_i + l_i \tag{23}$$

where x_{i+1}, x_i denotes the starting positions of the stage i, $i + 1$, and l_i the length between the position of stage i, $i + 1$.

Accepting distance as the discretization basis requires that velocity must be positive at every stage, which means choosing torque as the state variable will increase the computational complexity. As a result, we choose the speed as states.

FIGURE 2 The cost-to-achieve map for forward DP.

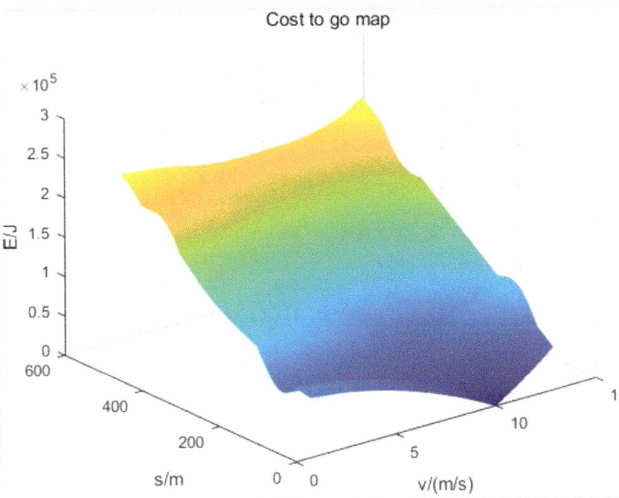

COST FUNCTION

Preventing the ego car from depending on sliding too much to save energy, we add the auxiliary cost and the energy to drive the wheels. Then the cost of every transition can be described as follows:

$$d_k = \frac{x_0}{\overline{v_k}}\left(P_{aux} + T_k \frac{i}{r_w} v_k\right) \tag{24}$$

where d_k denotes the transition energy cost in the stage number k, x_0 the distance we set in every stage, v_k the present velocity, P_{aux} the power that auxiliary equipment consume and T_k the torque we adopt in this stage. We assume that the process is a uniform acceleration or deceleration movement, then the average speed $\overline{v_k}$ is defined below:

$$\overline{v_k} = \frac{v_k + v_{k+1}}{2} \tag{25}$$

Therefore we have the optimal energy cost at stage k defined as

$$E = \sum_{k=1}^{n} d_k = \sum_{k=1}^{n} \frac{x_0}{\overline{v_k}}\left(P_{aux} + T_k \frac{i}{r_w} v_k\right) \tag{26}$$

If the velocity becomes v_{k+1}, according to (1)(2)(3)(4)(24):

$$v_{k+1} = v_k + \frac{x_0}{\overline{v_k}}\left[\frac{i\eta}{r_w m} T_k - \frac{\rho_d c_c A_f}{2m} v_k^2 - \left(c_r g \cos\alpha + g \sin\alpha\right)\right] \tag{27}$$

As a result, we can work out that the torque

$$T_k = \frac{\frac{v_{k+1}^2 - v_k^2}{2x_0} + \frac{\rho_d c_d A_f}{2m} v_k^2 + c_r g \cos\alpha + g \sin\alpha}{\frac{i\eta}{r_w m}} \tag{28}$$

Model Predictive Control (MPC)and Linear Quadratic Regulator (LQR)

LINEAR QUADRATIC REGULATOR (LQR)

LQR usually addresses optimal control problems with quadratic cost functions. For example,

$$\begin{aligned} x_{t+1} &= Ax_t + Bu_t, \\ s.t.x_0 &= x_{init} \end{aligned} \tag{29}$$

And

$$J = \min\left(\sum_{t=0}^{N-1} \left(x_t^{\,T} Q x_t + u_t^{\,T} R u_t \right) + x_N^{\,T} Q_f x_N \right) \tag{30}$$

with Q, R, Q_f as weight on x, u, and x_N, J as the cost of applying control inputs and being in the final state x_N.

In this situation, the problem is a nonlinear tracking problem. In fact, the cost function should be:

$$\begin{aligned} J = \min\Bigg(&\sum_{t=0}^{N-1} \left(\left(x_t - \overline{x_t}\right)^T Q \left(x_t - \overline{x_t}\right) \right) \\ &+ \sum_{t=0}^{N-1} \left(\left(u_t - \overline{u_t}\right)^T R \left(u_t - \overline{u_t}\right) \right) \\ &+ x_N^{\,T} Q_f x_N \Bigg) \end{aligned} \tag{31}$$

Second, considering the system as a nonlinear dynamic one, as is said in [17], we should start with a guess for U, and alternate between:

- linearize the problem around the current trajectory;
- solve associated LQR (tracking) problem.

In this situation, we set N as 2 to minimize the error caused by the process of linearization.

Meanwhile, tuning the parameters Q and R, the performance of the LQR tracking algorithm can be controlled. In other words, the output of the system can be regulated arbitrarily close to zero by adjusting the controllers' parameters.

MODEL PREDICTIVE CONTROL (MPC)

Model Predictive Control (MPC) is a form of control which is widely applied because of its ability of coping with hard constraints. It is able to optimize the control input of the present time slot while taking that of the future ones in a horizon into account. Linear MPC can achieve this with a given inner dynamic model and a stage cost J over the receding prediction horizon. At each time slot, it solves a Constrained Finite Time Optimal Control (CFTOC), and obtain a sequence of optimized control input. MPC apply it only during this timeslot, and solve a new CFTOC over a shifted horizon based on new situation.

In this case, MPC is adopted to track the given optimized speed trajectory while maintain security constraints. It addresses the optimal control problem with quadratic cost functions as below.

$$
\begin{aligned}
J = \min\Bigg(& \sum_{t=0}^{N-1}\left(\left(x_t - \overline{x_t}\right)^T Q\left(x_t - \overline{x_t}\right)\right) \\
& + \sum_{t=0}^{N-1}\left(\left(u_t - \overline{u_t}\right)^T R\left(u_t - \overline{u_t}\right)\right) \\
& + {x_N}^T Q_f x_N \Bigg) \\
s.t.\ \ & x_{k+1} = Ax_k + Bu_k, \\
& u_{\min} \le u_t \le u_{\max}, \\
& x_{\min} \le x \le x_{\max}.
\end{aligned}
\tag{32}
$$

with Q, R, Q_f as weight on x, the speed, u, the torque and x_N, the final speed in the horizon. $x_{k+1} = Ax_k + Bu_k$ is the given inner dynamic model while the J is the stage cost over the receding prediction horizon. Meanwhile, $u_{\min} \le u_t \le u_{\max}$ and $x_{\min} \le x \le x_{\max}$ serve as the constraints and N is the horizon.

Tuning the parameters Q, R and N, the performance of the MPC tracking algorithm can be controlled.

Simulation Results

In this session, we run some simulations on MATLAB to testif y the energ y saving when implementing the proposed strategy.

Unconstrained Speed Trajectories vs. Uniform Motion

First, we test the energy saving of unconstrained situation. For terrain information, we adopt the road slope map below which contains a hill and a valley.

The initial speed of the car in the optimal trajectory is set as 10 m/s. During calculations in DDP, the minimum and maximum of velocity choice in every stage are 1 m/s and 13 m/s respectively. Also we consider the feasible torque choice and the speed in the previous stage. Other cars running for comparisons are set to start with the initial speed of 6 m/s, 8 m/s, 9 m/s, 10 m/s, and 11 m/s, respectively and make uniform motion during the journey. The energy consumed are listed as follow.

According to Table 1, following the derived trajectory can save the energy needed to the degree of more than 27%. From the s-v image we can tell that compared to the uniform motion speed, 11 m/s is bigger than the highest speed in the unconstrained optimal trajectory, while 6 m/s is smaller than the lowest. However, the energy saving advantage remains. From the s-v chart, we can find that the optimal solution utilize the road slope by decelerating in the uphill and accelerating in the downhill. This is also in accord with usual human driving behavior.

Constrained vs. MPC and LQR

In the following part, trajectories derived from the algorithm mentioned above is compared with the route generated by the existing method of tracking problems with

FIGURE 3 The road slope map.

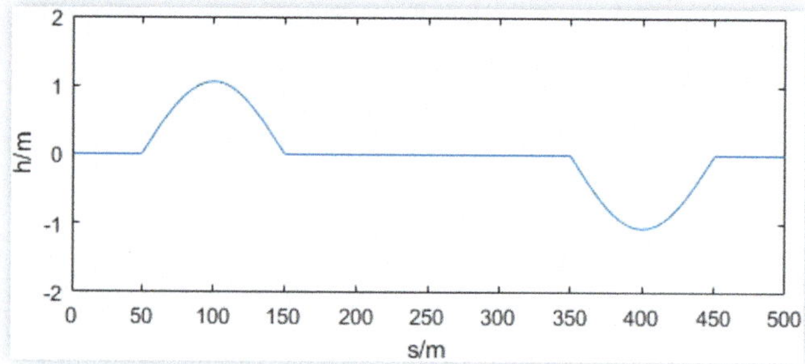

FIGURE 4 The optimal trajectory and cars making uniform motion.

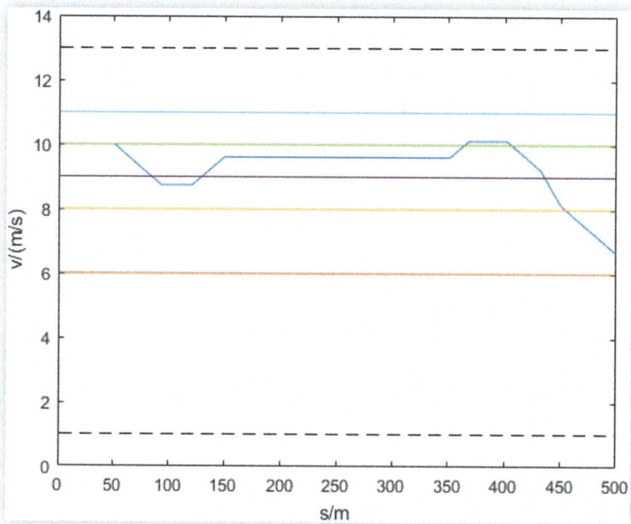

TABLE 1 Energy Consumed Table. Difference Column presents how much more energy that other cars consumed during the journey compared to cars following the unconstrained optimal speed trajectory

Process	Energy consumed/J	Difference/%
Displaced volume	224499.5725123820	0
Uniform Motion(v_0=6 m/s)	286259.1484299490	27.50988575
Uniform Motion(v_0=8 m/s)	286503.1638580620	27.6185788
Uniform Motion(v_0=9 m/s)	286655.8055436140	27.68657077
Uniform Motion(v_0=10 m/s)	286821.4745896320	27.76036559
Uniform Motion(v_0=11 m/s)	287004.3990123520	27.84184656

MPC. First we discuss the situation with one preceding platoon, then we consider up to three platoons.

ONE PRECEDING PLATOON

In this situation, we set the primary speed of the ego car as 8 m/s, the initial state of the preceding platoon to be 10 m, 7.5 m/s. Then we get the s-v chart and an energy consumed table as follows.

From Table 2 we can see that the constrained trajectory cost 2.50 kJ more than the unconstrained one. Meanwhile, its energy saving compared to the MPC tracking result

FIGURE 5 The s-v chart with one preceding platoon.

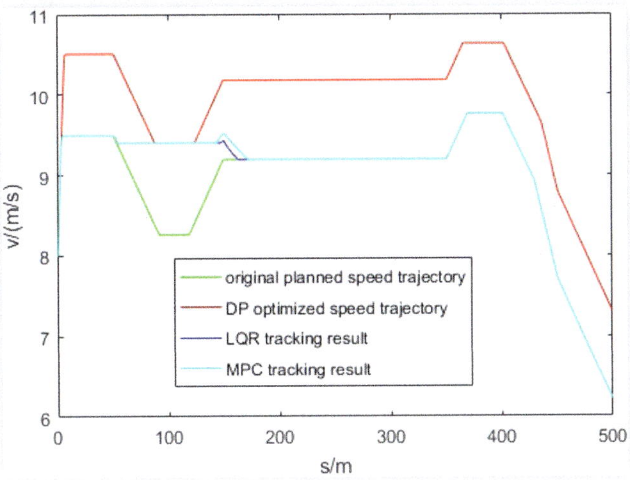

FIGURE 6 The s-Δx chart with one preceding platoon.

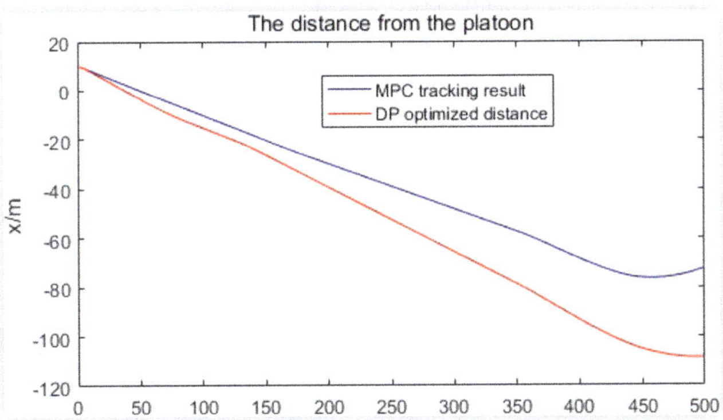

TABLE 2 Energy Consumed Table. Difference Column presents how much energy the constrained optimal trajectory can save with one platoon ahead

Process	Energy consumed/J	Difference/%
Unconstrained optimal trajectory	248890.68	
Constrained optimal trajectory	251392.78	0
MPC tracking trajectory	254291.30	1.15%
LQR tracking trajectory	260069.01	3.45%

is 1.15%, which compared to the LQR tracking result is 3.45%. Moreover, we see both of the trajectories overtaking at an early stage in Figure 6. In this case the optimal speed trajectory can be interpreted as accelerating to a certain speed (10.5 m/s) and practicing acceleration and deceleration as the optimal trajectory. Which means that utilizing road slope provides much more energy saving in this situation compared to air drag reduction.

TWO PRECEDING PLATOONS

In this part, we set the primary speed of the ego car 8 m/s, the initial state of the preceding platoon to be 10 m, 7 m/s and 35 m, 7.5 m/s. The simulation result is shown below.

Table 3 shows that the constrained optimal trajectory can save 1.99% energy of what the MPC tracking result need. According to Figure 7, we know that in the optimal speed

FIGURE 7 The s-v chart with two preceding platoons.

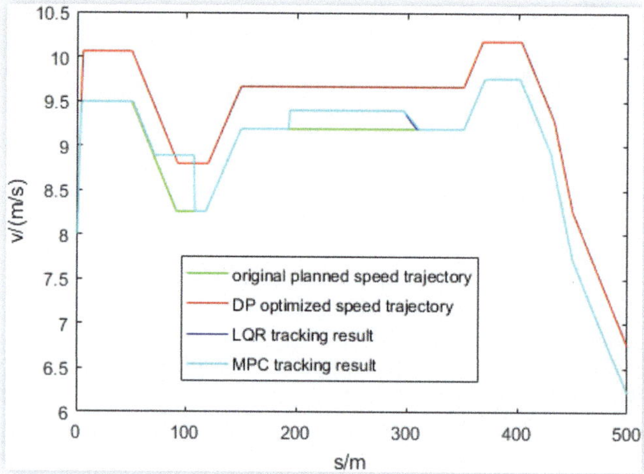

FIGURE 8 The s-Δx chart with two preceding platoons.

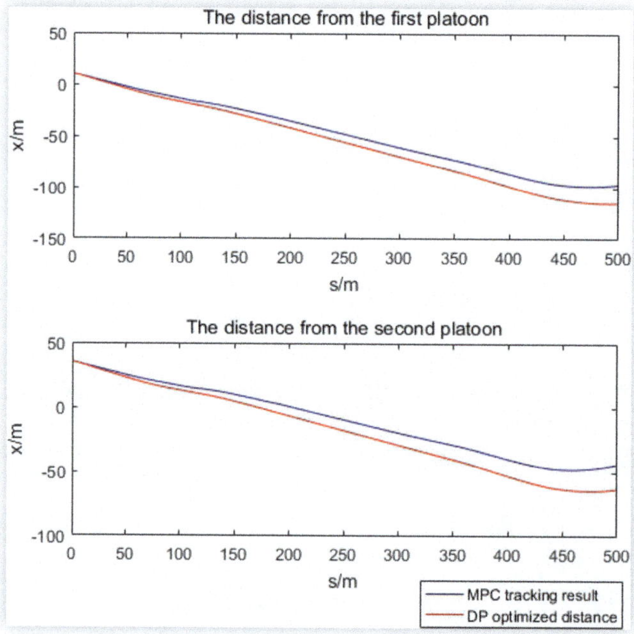

TABLE 3 Energy Consumed Table. Difference Column presents how much energy the constrained optimal trajectory can save with two preceding platoons

Process	Energy consumed/J	Difference/%
Unconstrained optimal trajectory	248890.68	
Constrained optimal trajectory	249545.65	0
MPC tracking trajectory	254511.74	1.99%
LQR tracking trajectory	257200.32	3.07%

trajectory, first the ego car accelerates to a certain speed (10.06 m/s) and then practices more deceleration (by 0.09 m/s) and less acceleration (by 0.15 m/s) than the optimal trajectory. In this way it does not need to adjust to the unconstrained one after overtaking as the MPC tacking solution do. Same as the former simulation, we can conclude that utilizing road slope provides much more energy saving in this situation compared to air drag reduction. Besides, instead of doing uniform motion in the last part of the road, it can also save a lot of energy from sliding there.

THREE PRECEDING PLATOONS

In this part, we set the primary speed of the ego car 8 m/s, the initial state of the preceding platoon to be 10 m, 7 m/s, 35 m, 7.5 m/s, and 57 m, 8.3 m/s. The simulation result is shown below.

Table 4 shows that constrained optimal trajectories can save 1.72% energy of the MPC tracking result. Figure 10 shows that the cars following the constrained optimal

FIGURE 9 The s-v chart with three preceding platoons.

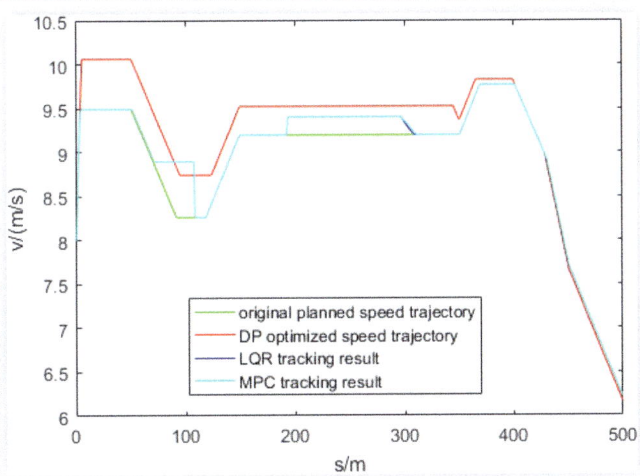

FIGURE 10 The s-Δxchart.

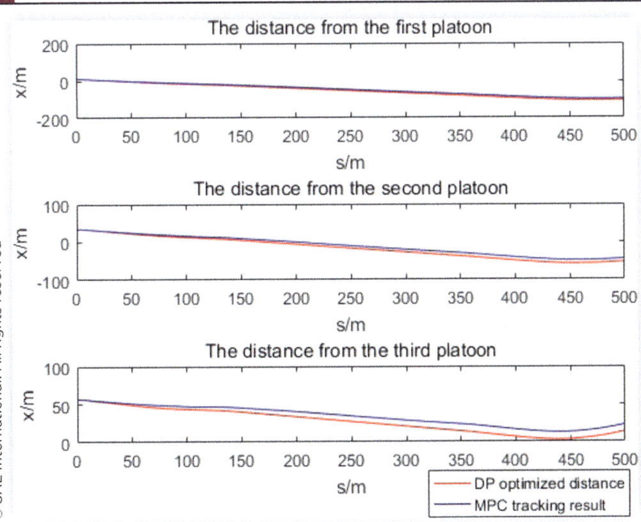

TABLE 4 Energy consumed table. Difference column presents how much energy the constrained optimal trajectory can save with three preceding platoons

Process	Energy consumed/J	Difference/%
Unconstrained optimal trajectory	248890.68	
Constrained optimal trajectory	250204.58	0
MPC tracking trajectory	254511.74	1.72%
LQR tracking trajectory	257200.32	2.80%

trajectory and the cars practicing MPC tracking overtake two platoons and merge into the third until the end. Figure 9 describes stages of the process. When going through the hill, the constrained optimal speed trajectory can be interpreted as following the same speed trajectory as that in the former scenario. Before it goes downhill in the valley($s \leq 350$m), it decelerates to increase the time following the third platoon while ensuring the satisfaction of safety constraints. This indicates that this algorithm be potent at handling the trade-off between the air drag reduction.

Sensitivity Analysis

Influence of Initial Speed on Optimal Trajectories

In this part, we set the primary state of the preceding platoons to be 10 m, 7 m/s, 35 m, 7.5 m/s, and 57 m, 8.3 m/s. Considering the influence of the initial speed of ego cars, we set it as 7 m/s, 8 m/s, 9 m/s, 9.25 m/s, 9.5 m/s, 10 m/s, 11 m/s. The simulation result is shown below.

Figure 11 shows that when the initial speed increases, the energy saving maintains at a high level (almost 5 kJ) and rises up to a peak of 19 kJ at the initial speed of 10 m/s.

FIGURE 11 The energy saving chart of constrained optimal trajectory compared to MPC tracking trajectory.

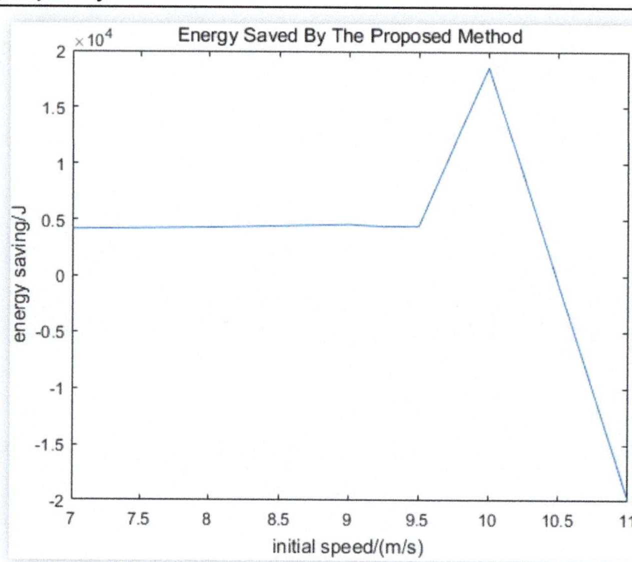

FIGURE 12 The instantaneous speed chart of trajectories which are developed by three approaches with three preceding platoons and different initial speed.

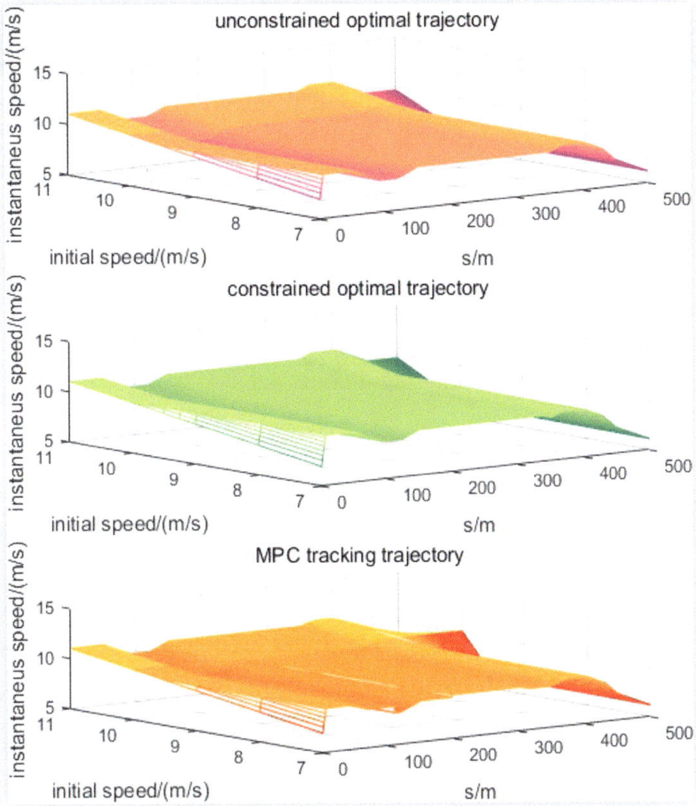

Accordingly, the proposed algorithm is potential for the urban environment and heavy duty trucks. When the initial speed is lower than 9.5 m/s, it is not sensitive to the change of initial speed. Figure 12 depicts that cars following constrained optimal trajectory slow down before the valley like the third simulation when the initial speed is relatively low (smaller than 11 m/s). Besides, cars follow similar trajectory when the initial speed is high (bigger than 10 m/s) and the energy saving effect is not significant. Basically, all cars are fast enough to overtake platoons without making up for unconstrained optimal trajectories. Meanwhile, the MPC results tend to either decelerate early (7 m/s, 8 m/s, 9 m/s, 9.25 m/s, and 9.5 m/s) to merge or to speed up and overtake(10 m/s) before the hill. The frequent acceleration and deceleration explains the energy wasted.

Conclusion

According to mentioned simulations, the proposed algorithm is able to derive optimal trajectories for ego cars with preceding platoons. It can utilize the road slope information after overtaking and air reduction when following platoons. This algorithm is especially potent for the cars in an urban environment or heavy duty trucks, which may be an advantage in the trunk transport of logistics industry.

For further study, three-dimensional grids of space may help consider lateral control of overtaking. Besides, a more sophisticated model with a mix of platoons and cars ahead can be a novel research project. Moreover, the setup of platoons can also be less stiff and researchers should consider interactions among the environment. For example, in the simulation with three preceding platoons, the deceleration before acceleration at around 350 meters from the starting place is harmless with the assumption that the ego car will not affect the control of platoons. In more practical scenerio, as is said in [18], we need to adjust the actuator to make sure that the fluctuation does not introduce instabilities to the control of platoons. In a word, this article introduces platoons to the optimization and does basic research for further research on optimal decisions in optimal trajectories.

Contact Information

Zheng Hongyu
State Key Laboratory of Automotive Simulation and
Control, Jilin University
Changchun, P. R. China
130025S
zhy_jlu@163.com

Acknowledgments

This work is supported by National Natural Science Foundation of China (51575223) and Key Laboratory for New Technology Application of Road Conveyance of Jiangsu Province (BM20082061712).

References

1. Transport Research & Information Portal, “Cooperative Intelligent Transport Systems (C-ITS),” accessed August 2017, http://www.transport-research.info/sites/default/files/TRIP_C-ITS_Report.pdf.
2. Chen, Y., Li, X., Wiet, C., and Wang, J., “Energy Management and Driving Strategy for In-Wheel Motor Electric Ground Vehicles with Terrain Profile Preview,” *IEEE Transactions on Industrial Informatics* 10, no. 3 (August 2014): 1938-1947, doi:10.1109/TII.2013.2290067.
3. Chen, Y. and Junmin, W., “Energy-Efficient Control Allocation for Over-Actuated Systems with Electric Vehicle Applications,” *Proceedings of the 2010 ASME Dynamic Systems and Control Conference*, 2010, doi:10.1115/DSCC2010-4012.
4. Chen, Y. and Wang, J., “Energy-Efficient Control Allocation with Applications on Planar Motion Control of Electric Ground Vehicles,” *Proceedings of the 2011 American Control Conference*, San Francisco, CA, 2011, 2719-2724, doi:10.1109/ACC.2011.5991160.
5. Chen, Y. and Wang, J., “Fast and Global Optimal Energy-Efficient Control Allocation with Applications to Over-Actuated Electric Ground Vehicles,” *IEEE Transactions on Control Systems Technology* 20, no. 5 (September 2012): 1202-1211, doi:10.1109/TCST.2011.2161989.

6. Li, S., Li, K., Rajamani, R., and Wang, J., "Model Predictive Multi-Objective Vehicular Adaptive Cruise Control," *IEEE Transactions on Control Systems Technology* 19, no. 3 (May 2011): 556-566, doi:10.1109/TCST.2010.2049203.
7. Li, S.E. and Peng, H., "Strategies to Minimize the Fuel Consumption of Passenger Cars during Car-Following Scenarios," *Proceedings of the Institution of Mechanical Engineers, Part D: Journal of Automobile Engineering* 226, no. 3 (2012): 419-429, doi:10.1177/0954407011420214.
8. Li, E., Shengbo, K.L., and Wang, J., "Economy-Oriented Vehicle Adaptive Cruise Control with Coordinating Multiple Objectives Function," *Vehicle System Dynamics* 51, no. 1 (2013): 1-17, doi:10.1080/00423114.2012.708421.
9. Li, S.E., Guo, Q., Xu, S., Duan, J. et al., "Performance Enhanced Predictive Control for Adaptive Cruise Control System Considering Road Elevation Information," *IEEE Transactions on Intelligent Vehicles* 99: 1-1, doi:10.1109/TIV.2017.2736246.
10. Liang, K.-Y., Mårtensson, J., and Johansson, K.H., "When Is it Fuel Efficient for a Heavy Duty Vehicle to Catch up with a Platoon?" *IFAC Proceedings Volumes* 46, no. 21 (2013): 738-743, doi:10.3182/20130904-4-JP-2042.00071.
11. Ajanović, Z., Stolz, M., and Horn, M., Energy Efficient Driving in Dynamic Environment: Considering Other Traffic Participants and Overtaking Possibility, *Comprehensive Energy Management-Eco Routing & Velocity Profiles* (Springer International Publishing, 2017), 61-80, ISBN:978-3-319-53164-9 (Print) 978-3-319-53165-6 (Online).
12. Rajamani, R., Tan, H.-S., Law, B.K., and Zhang, W.-B., "Demonstration of Integrated Longitudinal and Lateral Control for the Operation of Automated Vehicles in Platoons," *IEEE Transactions on Control Systems Technology* 8, no. 4 (July 2000): 695-708, doi:10.1109/87.852914.
13. Ajanović, Z., Stolz, M., and Horn, M., Energy-Efficient Driving in Dynamic Environment: Globally Optimal MPC-Like Motion Planning Framework, *Advanced Microsystems for Automotive Applications* (Cham: Springer, 2017), 111-122, doi:10.1007/978-3-319-66972-4_10.
14. Bergenhem, C., Shladover, S., Coelingh, E., Englund, C. et al., "Overview of Platooning Systems," *Proceedings of the 19th ITS World Congress*, Vienna, Austria, October 22-26, 2012.
15. Saerens, B., "Optimal Control Based Eco-Driving," PhD thesis, Department of Mechanical Engineering, Arenberg Doctoral School of Science, Engineering & Technology, Leuven, 2012.
16. Bellman, R., "The Theory of Dynamic Programming," No. RAND-P-550, Rand Corp, Santa Monica, CA, 1954.
17. Professor Stephen Boyd, "Lecture 1 Linear Quadratic Regulator: Discrete-Time Finite Horizon," EE 363, Winter 2008-09.
18. Dadras, S., Gerdes, R.M., and Sharma, R., "Vehicular Platooning in an Adversarial Environment," *Proceedings of the 10th ACM Symposium on Information, Computer and Communications Security*, ACM, 2015, 167-178, doi:10.1145/2714576.2714619.

epilogue

Juan R. Pimentel
Professor of Computer Engineering
Kettering University

In this book, we have characterized one of the main categories of automated vehicle safety, multi-agent safety. It is the safety category that most closely resembles the safety of vehicles when driven by a human; this means dealing with other vehicles, pedestrians, bicyclists, and any other object encountered while driving. There are a number of outstanding safety issues discussed in the book which will hopefully help clarify this topic, such as what is meant by multi-agent safety and how to measure the level of safety of automated vehicles, for example, whether the number (in millions) of miles driven is a good indicator. And if so, under what circumstances or conditions?

In addition to characterizing the safety category of multi-agent safety, we have included a set of ten papers which are representative of research in this category. Although there is a fair amount of work in this area, much of the work appears to be disjointed and performed in a somewhat ad hoc manner. One interesting point of view in the characterization of multi-agent safety is the model developed by researchers of reference [20], which has generated a great deal of discussion in the industry, particularly the concept of *responsibility-sensitive safety* (RSS), which involves the concept of blame, "typically only one of the agents is responsible for a\the accident, and therefore is to be blamed for it." As noted, much of the work on multi-agent safety is not integrated with the two other safety aspects, namely, functional safety and SOTIF. In this book, we have presented a risk-based model that integrates multi-agent safety into a risk-based approach. This requires additional definitions and/or more general interpretations of those in ISO 26262. The proposed model introduces two new sources of errors, failures, and hazards – *performance error* and *driving failure* – that model accident sources in the categories of SOTIF and multi-agent safety, respectively.

The topic of safety measures for measuring the level of safety is crucial for automated vehicle safety.

One type of safety measurement is around standards, processes, procedures, and design, and it is most useful at the design and simulation stages. One example of a safety measure that can be used during design is the number of expected catastrophic failures per year of operation (e.g., 500 hours) [24]. It can be argued that it would be difficult for all stakeholders to agree on a universal set of safety measures at this low level of abstraction. This fact is evident from the large amount of literature on the subject of multi-agent safety where many of the safety measures proposed are not universally discussed, accepted, or used by the research community. Another type of safety measurement at a much higher level that is useful to the general public and for developing appropriate public policy is that developed by the RAND Corporation [25], which consists of leading and lagging measures. The former reflect performance, activity, and prevention in the pre-crash phase, and the latter are observations of safety outcomes involving harm. Examples of leading measures can be proxy measures of driving behaviors correlated to safety outcomes. The RAND framework includes the following measures: infractions, roadmanship, disengagement (leading), and outcome (lagging) per development stages (development, demonstration, and deployment) and per test location (simulation, closed course, and public roads). Although useful for public policy, these safety measures do not help at the design stage where safety measures more closely related to design decisions are needed.